WORKSHEETS FOR CLASSROOM OR LAB PRACTICE

CHRISTINE VERITY

MATHEMATICS IN ACTION: AN INTRODUCTION TO ALGEBRAIC, GRAPHICAL, AND NUMERICAL PROBLEM SOLVING

FOURTH EDITION

The Consortium for Foundation Mathematics

Addison-Wesley
is an imprint of

Copyright © 2012 Pearson Education, Inc.
Publishing as Addison-Wesley, 75 Arlington Street, Boston, MA 02116.

ISBN-13: 978-0-321-73836-3
ISBN-10: 0-321-73836-5

1 2 3 4 5 6 BRR 14 13 12 11 10

Addison-Wesley
is an imprint of

www.pearsonhighered.com

Chapter 1 NUMBER SENSE

Activity 1.3

Learning Objectives
1. Identify and use the commutative property in calculations.
2. Use the distributive property to evaluate arithmetic expressions.
3. Use the order of operations convention to evaluate arithmetic expressions.
4. Identify and use the properties of exponents in calculations.
5. Convert numbers to and from scientific notation.
6. Identify, understand, and use formulas.
7. Use the order of operations convention in formulas involving whole numbers.

Key Terms
Use the vocabulary terms listed below to complete each statement in Exercises 1–5.

scientific notation	commutative	distributive
variable	formula	

1. The _____ property for addition can be written as $a + b = b + a$.

2. A number is expressed in _____ when it is written as the product of a decimal number between 1 and 10 and a power of 10.

3. A _____ shows the arithmetic relationship between two or more quantities.

4. The _____ property over addition can be written as $c(a + b) = c \cdot a + c \cdot b$.

5. A quantity that is represented by a letter or symbol is called a _____.

Practice Exercises
For #6-7, use the distributive property to evaluate each expression.

6. $16(60 - 3)$ 7. $(80 - 6) \cdot 7$ 6. _____

7. _____

For #8-15, evaluate each arithmetic expression by performing the operations in the appropriate order.

8. $\dfrac{9+11}{7-2}$

9. $54 \div (7+2)$

8. _____

9. _____

10. $98 - (63 - 21 \cdot 3) \div 6$

11. $4^3 \cdot 3^2 \div 16 - 11$

10. _____

11. _____

12. $9 \cdot 4^2 - 8 \cdot 3 + 7$

13. $385 \div 11 \cdot 6 - 12 \cdot 5$

12. _____

13. _____

14. $(5+3) \cdot 15 - 18 \div 3$

15. $\left(7^2 - 43\right)^2$

14. _____

15. _____

For #16-17, convert each number to scientific notation.

16. 376,050,000

17. 85,670,000,000

16. _____

17. _____

For #18-19, convert each number to standard notation.

18. 8.707×10^{9} **19.** 3.0499×10^{12}

18. _____

19. _____

For #20-21, perform each calculation.

20. $(5.32 \times 10^{5})(8.78 \times 10^{6})$ **21.** $\dfrac{8.10 \times 10^{18}}{4.5 \times 10^{11}}$

20. _____

21. _____

For #22-28, evaluate each formula for the given values.

22. $A = l \cdot w$, for $l = 45$ and $w = 8$ **23.** $A = \dfrac{1}{2}h(a+b)$, for $h = 6$, $a = 10$ and $b = 12$

22. _____

23. _____

24. $P = I \cdot V$, for $I = 12$ and $V = 115$ **25.** $P = 2l + 2w$, for $w = 6$ and $l = 10$

24. _____

25. _____

26. $I = P \cdot r \cdot t$, for   **27.** $V = \dfrac{1}{3}Bh$, for $B = 90$ and   **26.** _____

     $r = 0.04$, $P = 13{,}000$ $h = 15$

     and $t = 2$

27. _____

28. $A = 6s^3$, for $s = 4$ **28.** _____

Concept Connections

29. Melanie evaluates $3 \cdot (8 + 7) \cdot 5$ as $3 \cdot 8 + 7 \cdot 5$ by hand and gets 59 as the answer. What did she do wrong? What is the correct solution?

_____

_____

30. Beth writes 32,500,000,000 as 32.5×10^9. Explain why this is not in correct scientific notation. What is the correct answer?

_____

_____

Chapter 1 NUMBER SENSE

Activity 1.4

Learning Objectives
1. Add and subtract fractions.
2. Multiply and divide fractions.

Practice Exercises

For #1-28, perform the indicated operations.

1. $3\frac{1}{4}+5\frac{2}{3}$

2. $1\frac{3}{4}+6\frac{5}{7}$

3. $4\frac{1}{5}+\frac{17}{7}$

4. $\frac{12}{5}+2\frac{1}{3}$

5. $7\frac{2}{5}+3\frac{1}{6}$

6. $\frac{4}{3}+6\frac{7}{9}$

7. $3\frac{1}{3}+2\frac{1}{4}+1\frac{1}{2}$

8. $5\frac{3}{4}-1\frac{5}{7}$

1. _____

2. _____

3. _____

4. _____

5. _____

6. _____

7. _____

8. _____

9. $4\dfrac{1}{5} - 3\dfrac{2}{3}$

10. $9\dfrac{8}{9} - 2\dfrac{2}{3}$

9. _____

10. _____

11. $6\dfrac{3}{5} - 3\dfrac{5}{6}$

12. $10\dfrac{3}{4} - 8\dfrac{15}{16}$

11. _____

12. _____

13. $1\dfrac{3}{5} - 1\dfrac{1}{2}$

14. $7\dfrac{1}{6} - 4\dfrac{2}{5}$

13. _____

14. _____

15. $6\dfrac{2}{3} \cdot 3\dfrac{1}{6}$

16. $5\dfrac{1}{5} \cdot 7\dfrac{2}{7}$

15. _____

16. _____

17. $8\dfrac{1}{2} \cdot 10\dfrac{2}{5}$

18. $4\dfrac{1}{3} \cdot 9\dfrac{3}{4}$

17. _____

18. _____

Name: Date:

Instructor: Section:

19. $\dfrac{8}{3} \cdot \dfrac{3}{4}$

20. $\dfrac{7}{2} \cdot \dfrac{2}{3}$

19. _____

20. _____

21. $3\dfrac{3}{8} \cdot \dfrac{1}{4}$

22. $\dfrac{5}{4} \div \dfrac{7}{8}$

21. _____

22. _____

23. $2\dfrac{1}{3} \div 5$

24. $\dfrac{6}{7} \div \dfrac{2}{5}$

23. _____

24. _____

25. $1\dfrac{8}{9} \div \dfrac{4}{7}$

26. $3\dfrac{1}{5} \div \dfrac{10}{7}$

25. _____

26. _____

27. $5 \div \dfrac{3}{4}$

28. $8\dfrac{1}{3} \div 2\dfrac{2}{5}$

27. _____

28. _____

Concept Connections

29. Alice incorrectly simplifies $\dfrac{2}{3} + \dfrac{7}{9}$ as $\dfrac{9}{9} = 1$. What did she do wrong? What is the correct solution?

_____

_____

30. Roy incorrectly simplifies $\dfrac{3}{8} \div \dfrac{4}{7}$ as $\dfrac{3}{2} \div \dfrac{1}{7} = \dfrac{3}{2} \cdot \dfrac{7}{1} = \dfrac{21}{2}$. What did he do wrong? What is the correct solution?

_____

_____

Chapter 1 NUMBER SENSE

Activity 1.5

Learning Objectives
1. Recognize and calculate a weighted average.

Key Terms
Use the vocabulary terms listed below to complete each statement in Exercises 1–2.

simple average   **weighted average**

1. To calculate a _____, multiply each data value by the ratio of importance and sum.
2. To calculate a _____, add all the scores and divide the sum by the number of exams.

Practice Exercises
For #3-10, find the average of the following test scores.
Round to the nearest point.

3. 83, 76, 86, 94   4. 74, 61, 79, 68, 70   3. _____

                                                                 4. _____

5. 96, 94, 89, 83, 79   6. 67, 87, 91, 89, 75   5. _____

                                                                 6. _____

7. 75, 71, 69, 58, 66, 81   8. 59, 67, 73, 78, 65   7. _____

                                                                 8. _____

9. 85, 82, 89, 73, 96   **10.** 72, 86, 70, 84, 74

9. _____

10. _____

For #11-16, find the average of the following quiz scores.
Round to the nearest point.

11. 9, 8, 10, 7, 6   **12.** 10, 9, 10, 5, 6

11. _____

12. _____

13. 8, 8, 7, 7, 6, 10, 3   **14.** 10, 5, 9, 9, 8, 3

13. _____

14. _____

15. 5, 6, 5, 10, 6, 7, 8, 10, 0   **16.** 7, 8, 7, 9, 8, 7, 9

15. _____

16. _____

For #17-24, find each student's grade using weighted averages. The quiz score represents $\frac{1}{4}$ of the grade, the homework score represents $\frac{1}{5}$ of the grade, the test score represents $\frac{1}{2}$ of the grade, and class participation represents $\frac{1}{20}$ of the grade. Round to the nearest point.

17. Quiz: 90
Homework: 85
Test: 80
Participation: 70

18. Quiz: 73
Homework: 90
Test: 85
Participation: 86

17. _____

18. _____

19. Quiz: 64
Homework: 76
Test: 72
Participation: 70

20. Quiz: 96
Homework: 94
Test: 86
Participation: 98

19. _____

20. _____

21. Quiz: 83
Homework: 60
Test: 75
Participation: 81

22. Quiz: 89
Homework: 98
Test: 82
Participation: 96

21. _____

22. _____

23. Quiz: 78
Homework: 66
Test: 85
Participation: 75

24. Quiz: 85
Homework: 85
Test: 85
Participation: 85

23. _____

24. _____

For #25-28, find each battering average. Round to the nearest thousandth.

25. Hits: 18
At bats: 52

26. Hits: 17
At bats: 60

25. _____

26. _____

27. Hits: 13
At bats: 31

28. Hits: 27
At bats: 75

27. _____

28. _____

Concept Connections

29. What is the weight of each score of 5 exams that have equal weights?

_____

_____

30. When computing weighted averages, what can be said about the sum of the weights?

_____

_____

Chapter 1 NUMBER SENSE

Activity 1.6

Learning Objectives
1. Distinguish between absolute and relative measure.
2. Write ratios in fraction, decimal, and percent formats.
3. Determine equivalence of ratios.

Key Terms
Use the vocabulary terms listed below to complete each statement in Exercises 1–6.

percent	ratio	relative measure
verbal	fractional	decimal

1. _____ is a word used to describe the relative measure quotient.

2. A ratio can be expressed in the form _____ $\left(\dfrac{4}{5}\right)$.

3. A ratio can be expressed in the form _____ (4 out of 5).

4. A ratio can be expressed in the form _____ (0.8).

5. _____ is a quotient that compares two similar quantities, often a "part" and a "total." The part is divided by the total.

6. _____ always indicates a ratio out of 100.

Practice Exercises
For #7–12, write the ratio as a fraction.

7. 55 out of 103   8. 32 out of 99   7. _____

8. _____

9. 13 out of 20   10. 5 out of 9   9. _____

10. _____

11. 99 out of 100   **12.** 6 out of 7

11. _____

12. _____

For #13–18, write the fraction as a decimal.

13. $\dfrac{17}{20}$ **14.** $\dfrac{35}{80}$

13. _____

14. _____

15. $\dfrac{24}{96}$ **16.** $\dfrac{27}{70}$

15. _____

16. _____

17. $\dfrac{42}{90}$ **18.** $\dfrac{39}{52}$

17. _____

18. _____

For #19–23, write the ratio in percent format.
19. 37 out of 50   **20.** 9 out of 25

19. _____

20. _____

21. 7 out of 16   **22.** 9 out of 30   **21.** _____

                                                             22. _____

23. 9 out of 72   **23.** _____

For #24–28, write the percent in decimal format.

24. 83%   **25.** 4.5%   **24.** _____

                                                             25. _____

26. 350%   **27.** 0.63%   **26.** _____

                                                             27. _____

28. 1.23%   **28.** _____

Concept Connections

29. After a basketball game, Tyus said he made 12 baskets, and Jayson said he made 16 baskets. You want to determine the relative performance for each player. What information is missing?

_____

_____

30. From Exercise #29, suppose that Tyus had made 18 attempts, and Jayson had 22 attempts. Which player can claim a better performance?

_____

_____

Chapter 1 NUMBER SENSE

Activity 1.7

Learning Objectives
1. Use proportional reasoning to apply a known ratio to a given piece of information.

Key Terms
Use the vocabulary terms listed below to complete each statement in Exercises 1–3.

| proportional reasoning | multiplied by | divided by |

1. If the given information represents a total, then the part is unknown:
   unknown total = part _____ ratio.

2. If the given information represents a part, then the total is unknown:
   unknown part = total _____ ratio.

3. _____ is the process by which you apply a known ratio to one piece of information to determine a related piece of information.

Practice Exercises
For 4–27, determine the value of each expression.

4. $15 \cdot \dfrac{4}{5}$ 5. $24 \div \dfrac{3}{4}$ **4.** _____

                                                                5. _____

6. $\dfrac{1}{6}$ of 72   7. $\dfrac{2}{3}$ of 36   **6.** _____

                                                                7. _____

8. $32 \div \dfrac{4}{5}$

9. $72 \div \dfrac{8}{9}$

8. _____

9. _____

10. 20% of 70

11. 25% of 32

10. __19_____

11. _____

12. $120 \div 0.80$

13. $240 \div 0.30$

12. _____

13. _____

14. $35 \cdot \dfrac{4}{7}$

15. $48 \cdot \dfrac{5}{6}$

14. _____

15. _____

16. 35% of 840

17. 72% of 450

16. _____

17. _____

18. $960 \div 80\%$

19. $8550 \div 45\%$

18. _____

19. _____

20. $3000 \div \dfrac{8}{9}$

21. $1000 \div \dfrac{5}{7}$

20. _____

21. _____

22. 90% of 330

23. 23% of 420

22. _____

23. _____

24. $20{,}000 \div 40\%$

25. $3500 \div 20\%$

24. _____

25. _____

26. 19% of 650

27. $762 \div 25\%$

26. _____

27. _____

Concept Connections

28. There are 480 children at a local elementary school. If 15% of the children are left-handed, find the number of children that are left-handed.

_____

_____

29. From Exercise, #28, assuming that the children at the local elementary school are either left-handed or right-handed, how many children are right-handed?

_____

_____

30. From Exercise #29, what percent of children at the local elementary school are right-handed?

_____

_____

Chapter 1 NUMBER SENSE

Activity 1.8

Learning Objectives
1. Define actual and relative change.
2. Distinguish between actual and relative change.
3. Calculate relative change as a percent increase or percent decrease.

Key Terms
Use the vocabulary terms listed below to complete the equation for Exercises 1–2.

actual change   **relative change**

$$(1)\underline{\hspace{4cm}} = \frac{(2)\underline{\hspace{3cm}}}{\text{original value}}$$

Practice Exercises
For 3–27, solve.

3. Last year, the Lab fee for a course was $50. This year the Lab fee for the same course is $65. Determine the increase in the Lab fee.

3. _____

4. From Exercise #3, determine the relative increase in the Lab fee.

4. _____

5. Last year, the textbook for a statistics class was $144. This year the textbook for the same class is $168. Determine the increase in price.

5. _____

6. From Exercise #5, determine the relative increase in price.

6. _____

7. Last year, the textbook for a history class was $95. This year the textbook for the same class is $115. Determine the increase in price.

7. _____

8. From Exercise #7, determine the relative increase in price.

8. _____

9. From Exercises #5 and #7, which textbook had the larger percentage increase?

9. _____

10. The fall semester had an enrollment of 5200 students. The spring semester there were only 5035 students. Determine the decrease in enrollment.

10. _____

11. From Exercise #10, determine the relative decrease (as a percent) in enrollment.

11. _____

12. The first summer session had an enrollment of 1520.   **12.** _____
The second summer session had only 925 students.
Determine the decrease in enrollment.

13. From Exercise #12, determine the relative decrease   **13.** _____
(as a percent) in enrollment.

14. From Exercises #10 and #12, which semester or   **14.** _____
session had the largest decrease in enrollment?

15. Before Tyler joined a gym, he weighed 247 lbs. After   **15.** _____
a year in the exercise program, he weighed 212 lbs.
Determine the decrease in weight.

16. From Exercise #15, determine the relative decrease   **16.** _____
(as a percent) in weight.

17. Tyler quit the gym when he weighed 212 lbs and gained weight. He now weighs 247 lbs. Determine the increase in weight.

17. _____

18. From Exercise #17, determine the relative increase (as a percent) in weight.

18. _____

19. A prized rose bush has an average bloom of 15. Using a new fertilizer the number of blooms increases to 21. Determine the increase in blooms.

19. _____

20. From Exercise #19, determine the relative increase (as a percent) in blooms.

20. _____

21. A prized fertilized rose bush has an average bloom of 21. Lack of water decreases the number of blooms to 8. Determine the decrease in blooms.

21. _____

22. From Exercise #21, determine the relative decrease (as a percent) in blooms.

22. _____

23. From Exercises #19 and #21, does fertilizer or lack of water cause the largest change in blooms?

23. _____

24. The average price of a gallon of gasoline was $2.77. During the vacation season a gallon of gasoline was $2.95. Determine the actual increase in the price of gasoline.

24. _____

25. From Exercise #24, determine the relative increase (as a percent) in the price.

25. _____

26. After the vacation season, the price of a gallon of gasoline drops to $2.77 from $2.95. Determine the actual decrease in the price of gas.

26. _____

27. From Exercise #26, determine the relative decrease (as a percent) in the price.

27. _____

Concept Connections

28. You buy stock with a company for $50. The stock price increases to $150. Find the percent increase. Why is this percent more than 100%?

_____

_____

29. You buy stock with a company for $100. The stock price decreases to $25. Find the percent decrease. When dealing with the stock market, will a percent decrease ever be less than 0%?

_____

_____

30. You buy stock with a company for $58. Under what circumstances would the percent decrease reach 100%?

_____

_____

Chapter 1 NUMBER SENSE

Activity 1.9

Learning Objectives
1. Define growth factor.
2. Determine growth factors from percent increases.
3. Apply growth factors to problems involving percent increases.
4. Define decay factor.
5. Determine decay factors from percent decreases.
6. Apply decay factors to problems involving percent decreases.

Key Terms
Use the vocabulary terms listed below to complete each statement in Exercises 1–2.

> **growth factor**   **decay factor**

1. When a quantity decreases by a specified percent, the ratio of its new value to the original value is called the _____ .

2. When a quantity increases by a specified percent, the ratio of its new value to the original value is called the _____ .

Practice Exercises
For 3–27, solve.

3. You are purchasing a new car. The purchase price of the car in Sacramento, California is $21,995 excluding sales tax. The sales tax rate is 8.75%. What growth factor is associated with the sales tax rate?

    3. _____

4. From Exercise #3, use the growth factor to determine the total cost of the car.

    4. _____

5. The cost of bus fare in a small city is now $2. Last June the price was increased by 14.3%. What was the pre-June price of the fare? Round to the nearest penny.

5. _____

6. One year, a city high school had enrollment of 1195 students. The next year enrollment increased to 1244. What is the ratio of the enrollment in the second year to the enrollment in the first year?

6. _____

7. From Exercise #6, determine the growth factor. Write in decimal form to the nearest thousandth.

7. _____

8. From Exercise #6, by what percent did the enrollment increase?

8. _____

9. The number of registered voters in a city in Oregon increased by 13.5% from 2000 to 2008. In 2008, the number of registered voters was reported to be 27,919. What was the number of registered voters in the Oregon city in 2000?

9. _____

10. In 1979, your grandparents bought a house for $50,000. Due to inflation, the cost of the same house in 2009 was $157,500. What is the inflation growth factor of housing from 1979 to 2009?

10. _____

11. From Exercise #10, what is the inflation rate of housing from 1979 to 2009?

11. _____

12. From Exercise #10, the house was sold in 2009 for $189,000. What profit was made in terms of 2009 dollars?

12. _____

13. Your friend plans to move from Alabama to Connecticut. He earns $40,000 a year in Alabama. How much must he earn in Connecticut to maintain the same standard of living if the cost of living in Connecticut is 38% higher?

13. _____

14. A one-year certificate of deposit earns an annual percentage yield of 4.05%. If $4000 is invested, how much will the investment be worth in a year?

14. _____

15. You wrote a 24-page booklet to be published. The editor asked you to reduce the number of pages to 20. By what percent must you reduce the length of the booklet?

15. _____

16. In 1998 in a city in Ohio, the number of children enrolled in the city preschool program was 20,162. In 2008, the number of children enrolled in the city preschool program dropped to 19,073. For the 10-year period, what is the decay factor for the city preschool program?

16. _____

17. From Exercise #16, what is the percent decrease?

17. _____

18. A deluxe paper shredder has a list price of $149.99. On Sunday, the model goes on sale and the price is reduced by 20%. What is the decay factor?

18. _____

19. From Exercise #18, use the decay factor to determine the sale price of the deluxe paper shredder.

19. _____

20. You are on a weight-reducing diet. Your goal is to lose   **20.** _____
15% of your body weight over the next year. You
currently weigh 235 pounds. What is the decay factor?

21. From Exercise #19, use the decay factor to calculate the   **21.** _____
goal weight to the nearest pound.

22. After one year, you reach your goal weight. However, you   **22.** _____
have more weight to lose to reach your ideal weight range,
which is 170-185 pounds. If you lose 8% of your new
weight, will you be in your ideal weight range? Explain.

_____

_____

23. In general, 75% of a dose of aspirin is eliminated from the   **23.** _____
blood stream in an hour. A person takes a low dose of
aspirin of 81 milligrams. How much aspirin remains in the
person after 1 hour?

24. You are flying from Washington D.C. to Chicago. The   **24.** _____
restricted, nonrefundable fare is $176.40. You book online
to take advantage of the 5% discount on the ticket. What is
the decay factor?

25. From Exercise #24, determine the cost of the fare if you book online.

25. _____

26. A deluxe wireless router has a list price of $69.99. Yesterday, the model went on sale and the price was reduced by 25%. What is the decay factor?

26. _____

27. From Exercise #26, use the decay factor to determine the sale price of the deluxe wireless router.

27. _____

Concept Connections

28. After applying a growth factor, which is larger, the original value or the new value? Why?

_____

_____

29. After applying a decay factor, which is larger, the original value or the new value? Why?

_____

_____

30. How would you describe a factor that is equal to 1?

_____

_____

Chapter 1 NUMBER SENSE

Activity 1.10

Learning Objectives
1. Define consecutive growth and decay factors.
2. Determine a consecutive growth or decay factor from two or more consecutive percent changes.
3. Apply consecutive growth and/or decay factors to solve problems involving percent changes.

Key Terms

Choose the correct term listed in parentheses to complete each statement in Exercises 1–2.

1. The cumulative effect of a sequence of percent changes is the
_____ (product/quotient) of the associated growth or decay factors.

2. With the cumulative effect of a sequence, the order in which the changes are applied
_____ (does/does not) matter.

Practice Exercises

For 3–28, solve.

3. A $150 leather purse is on sale for 25% off. Determine the   **3.** _____
decay factor for the sale.

4. From Exercise #3, you present a coupon for an additional   **4.** _____
15% off. Determine the decay factor for the additional
discount.

5. From Exercises #3 and #4, use these decay factors to   **5.** _____
determine the price you pay for the purse.

6. A pair of jeans has a regular price of $79.99. The sale is for 35% off. Determine the decay factor for the sale.

6. _____

7. From Exercise #6, you present a coupon for an additional 10% off. Determine the decay factor for the additional discount.

7. _____

8. From Exercises #6 and #7, use the decay factors to determine the price you pay for the jeans.

8. _____

9. Your union has negotiated a 2-years contract containing annual raises of 3.5% and 4.5% during the term of the contract. Your current salary is $52,500. What salary will you earn in 2 years?

9. _____

10. Your union has negotiated a 3-year contract containing annual raises of 2.5%, 3%, and 3.5% during the term of the contract. Your current salary is $49,000. What salary will you earn in 3 years?

10. _____

11. As a vendor at an open air market, you increased your inventory of veggie yogurt cheese of 115 pounds by 15%. You sold 85% of your inventory. How many pounds of cheese remain?

11. _____

12. As the owner of a news stand, you anticipate a large demand for a popular sports magazine and increase your inventory of 250 by 20%. You sold 70% of your inventory. How many sports magazines remain?

12. _____

13. You deposit $2500 in a 3-year certificate of deposit that pays 3.5% interest compounded annually. Determine, to the nearest dollar, your account balance when your certificate comes due.

13. _____

14. You deposit $3500 in a 4-year certificate of deposit that pays 3% interest compounded annually. Determine, to the nearest dollar, your account balance when your certificate comes due.

14. _____

15. Determine a single decay factor that represents the cumulative effect of consecutively applying discounts of 50% and 45%.

15. _____

16. From Exercise #15, use the decay factor to determine the effective discount.

16. _____

17. Determine a single decay factor that represents the cumulative effect of consecutively applying discounts of 30%, 40%, and 50%.

17. _____

18. From Exercise #17, use this decay factor to determine the effective discount.

18. _____

19. A charitable organization has decreased its budget of $450,000 by 6% in each of the last 3 years. What is the current budget?

19. _____

20. A small church has decreased its budget of $59,000 by 4.5% in each of the last 4 years. What is the current budget?

20. _____

21. An appliance store has decreased its budget of $117,600 for advertising by 8% in each of the last 4 years. What is the current budget?

21. _____

22. A high growth stock is purchased for $2500. The value rose by 25%. Determine the growth factor.

22. _____

23. After a year, the value of a high growth stock from Exercise #22 drops by 25%. Determine the decay factor.

23. _____

24. From Exercises #22 and #23, what is the single factor that represents the increase and decrease?

24. _____

25. From Exercise #24, does the number represent a growth factor or a decay factor? Why?

25. _____

26. From Exercises #22 and #23, what is the current value of the high growth stock?

26. _____

27. From Exercises #22 – 26, what is the cumulative effect of applying a 25% increase followed by a 25% decrease?

27. _____

28. From Exercises #22 – 26, what is the cumulative effect if the 25% decrease had been applied first, followed by the 25% increase?

28. _____

Concept Connections

29. Which is a better deal: 15% discount and additional 40% discount, or 20% discount and additional 30% discount?

30. Which is a better deal: 30% discount or 10% discount and additional 20% discount?

Chapter 1 NUMBER SENSE

Activity 1.11

Learning Objectives
1. Apply rates directly to solve problems.
2. Use unit or dimensional analysis to solve problems that involve consecutive rates.

Practice Exercises

For 1–28, convert.

1. Convert 15 hours to seconds.   1. _____

2. Convert 2 meters to inches.   2. _____

3. Convert 4 km to miles.   3. _____

4. Convert 80 km/hr to mph.   4. _____

5. Convert 180 cm to inches.   5. _____

6. Convert 33,840 seconds to hours.　　　　　　**6.** _____

7. Convert 2.5 weeks to minutes.　　　　　　**7.** _____

8. Convert 3L/day to milliliters/hour.　　　　　　**8.** _____

9. Convert 48 mg/day to grams in 30 days.　　　　　　**9.** _____

10. Convert 2.4L/day to milliliters per hour.　　　　　　**10.** _____

11. Convert 162 inches to yards.　　　　　　**11.** _____

Name: Date:
Instructor: Section:

12. Convert 2640 feet to miles.   **12.** _____

13. Convert 3 miles to inches.   **13.** _____

14. Convert 1.5 yards to meters   **14.** _____

15. Convert 35 mg/week to grams/12 weeks   **15.** _____

16. Convert 5 weeks to minutes.   **16.** _____

17. A marathon race is 26.2 miles. Find the distance in feet. **17.** _____

18. What is the distance of the marathon in kilometers? **18.** _____

19. What is the distance of the marathon in meters? **19.** _____

20. Your car's highway fuel efficiency is 34 miles per gallon. What is its fuel efficiency in kilometers per liter? **20.** _____

21. Your SUV's highway fuel efficiency is 25 miles per gallon. What is its fuel efficiency in kilometers per liter? **21.** _____

22. Your recreational vehicle's highway fuel efficiency is 12 miles per gallon. What is its efficiency in kilometers per liter? **22.** _____

23. A soda bottle contains 2 liters of cola. How many   **23.** _____
quarts of cola are in the soda bottle?

24. How many pints of cola are in a 2 liter bottle of cola?   **24.** _____

25. A sprinter runs the 200 yard dash. What is the length   **25.** _____
in feet?

26. How many seconds are there in 3 days?   **26.** _____

27. How many hours are in 3 weeks?   **27.** _____

28. How many pints are in 3 liters?   **28.** _____

Concept Connections

29. In the movie Julie and Julia, Julia states that when writing the American cookbook, one difficulty she had with replicating French recipes, was getting the amounts correct. What was she talking about?

_____

_____

30. At the grocery store, many packaged products contain more than one unit of measure. For example, a leading brand of peanut butter claims to contain 18 oz, or 510 g. How can you use this information to determine a rate of unit analysis to convert between ounces and grams?

_____

_____

Chapter 1 NUMBER SENSE

Activity 1.12

Learning Objectives
1. Identify signed numbers.
2. Use signed numbers to represent quantities in real-world situations.
3. Compare signed numbers.
4. Calculate the absolute value of numbers.
5. Identify and use properties of addition and subtraction of signed numbers.
6. Add and subtract signed numbers using absolute value.

Practice Exercises
For #1-24, evaluate each expression.

1. $-4+(-11)$ **2.** $-7+3$ **1.** _____

2. _____

3. $5+(-17)$ **4.** $19-7$ **3.** _____

4. _____

5. $8-15$ **6.** $-3-11$ **5.** _____

6. _____

7. $-8+(-11)$ **8.** $(-8)+12$ **7.** _____

8. _____

9. $-7-11$

10. $-6-(-7)$

9. _____

10. _____

11. $0-(-9)$

12. $-5+0$

11. _____

12. _____

13. $18-18$

14. $18-(-18)$

13. _____

14. _____

15. $0+(-11)$

16. $-5+(-6)$

15. _____

16. _____

17. $-9-(-4)$

18. $-8-(-13)$

17. _____

18. _____

19. $7-10$

20. $(-11)-5$

19. _____

20. _____

Name: Date:
Instructor: Section:

21. $-\dfrac{6}{7}+\dfrac{2}{7}$

22. $6.7-7.3$

21. _____

22. _____

23. $|-15|$

24. $\left|-3\dfrac{1}{3}\right|$

23. _____

24. _____

For #25-28, insert the appropriate inequality symbol.

25. $-6\ \square\ -3$

26. $-9\ \square\ 0$

25. _____

26. _____

27. $-4\ \square\ -20$

28. $6\ \square\ -5$

27. _____

28. _____

Concept Connections

29. Most competitive professional sports have scores that are zero, or positive numbers. Name a sport that uses negative numbers throughout the scoring process and state how the negative numbers are used to represent the score.

_____

_____

30. You look at a topological map of California. You notice that the highest point in California is Mt. Whitney at 14,494 ft and the lowest point is in Death Valley at −282 ft. What do the positive and negative numbers mean for these places?

_____

_____

Chapter 1 NUMBER SENSE

Activity 1.13

Learning Objectives
1.   Multiply and divide signed numbers.

Practice Exercises

For #1-28, determine the product or quotient.

1. $-6(-8)$ **2.** $(-9)(-7)$

3. $-7 \cdot 3$ **4.** $(-5)(-8)$

5. $-1(3.14)$ **6.** $-7.389(-1)$

7. $-11(0)$ **8.** $5(-6)$

1. _____

2. _____

3. _____

4. _____

5. _____

6. _____

7. _____

8. _____

9. $(-4)(-7)$ **10.** $63 \div (-9)$ **9.** _____

10. _____

11. $-35 \div (-7)$ **12.** $0 \div (-6)$ **11.** _____

12. _____

13. $-14 \div 0$ **14.** $-24 \div 6$ **13.** _____

14. _____

15. $18 \div (-3)$ **16.** $-54 \div 9$ **15.** _____

16. _____

17. $-72 \div (-1)$ **18.** $-3 \div 12$ **17.** _____

18. _____

19. $\dfrac{1}{4} \div (-4)$ **20.** $-4 \div \dfrac{1}{4}$ **19.** _____

20. _____

21. $(3.1)(-4)$

22. $2(-11)(-5)$

21. _____

22. _____

23. $0.4(-0.5)$

24. $(-7)(-8)(-1)(4)(3)$

23. _____

24. _____

25. $\left(-\dfrac{2}{3}\right)\left(-\dfrac{9}{10}\right)$

26. $-6\dfrac{3}{4} \div \left(-\dfrac{9}{10}\right)$

25. _____

26. _____

27. $-27 \div 0$

28. $0 \div (-50)$

27. _____

28. _____

Concept Connections

29. After a series of rainstorms, a river crested to 30 ft, near flood stage. After 6 days, the river was marked at 24 ft. What was the daily average change of the river over the 6 day period?

_____

_____

30. Symbolically, write a quantity that is equivalent to negative one-half, three different ways, changing only the location of the negative symbol.

_____

_____

Chapter 1 NUMBER SENSE

Activity 1.14

Learning Objectives
1. Use the order of operations convention to evaluate expressions involving signed numbers.
2. Evaluate expressions that involve negative exponents.
3. Distinguish between such expressions as -5^4 and $(-5)^4$.
4. Write very small numbers in scientific notation.

Practice Exercises
For #1-28, evaluate each expression.

1. $(7-15) \div 4 + 9$

2. $-6 + 5(2 - 7)$

3. $-4^2 \cdot 3^2 + 64$

4. $1.5 - (2.1 + 3.9)^2$

5. $\dfrac{4}{15} - 5\left(\dfrac{8}{15} - \dfrac{11}{15}\right)$

6. $\dfrac{2}{3} \div \left(\dfrac{3}{4} - \dfrac{11}{12}\right)$

7. -1^{-4}

8. $5^3 - (-5)^3$

1. _____

2. _____

3. _____

4. _____

5. _____

6. _____

7. _____

8. _____

9. $(13)^2 - (13)^2$

10. $\dfrac{-6 \times 10^3}{\left(2 \times 10^{-2}\right) - \left(3 \times 10^{-2}\right)}$

9. _____

10. _____

11. $\dfrac{2.5 \times 10^4}{\left(-4.0 \times 10^{-2}\right)\left(3.2 \times 10^3\right)}$

12. 8^{-2}

11. _____

12. _____

13. $(-8)^2$

14. -8^2

13. _____

14. _____

15. $(-8)^{-2}$

16. $\dfrac{7}{8} \div (-14) + \dfrac{5}{16}$

15. _____

16. _____

17. $(3-8)5$

18. $(3+4)^2 - 25$

17. _____

18. _____

19. $-9 + 12 \div (8 - 10)$

20. $3.5 - (4.3 - 1.3)^2 + 6$

19. _____

20. _____

21. $\dfrac{1}{6} - \left(\dfrac{3}{4} \cdot \dfrac{2}{9} \right)$

22. $5.4 \div 0.9 - (-3.6 + 5.3)$

21. _____

22. _____

23. $(3 - 5)^2$

24. $-\left(\dfrac{1}{3} \right)^3 + \left(\dfrac{1}{3} \right)^3$

23. _____

24. _____

25. $5 - 3^2$

26. $(-1)^{10}$

25. _____

26. _____

27. $4.2 \cdot 10^{-3}$

28. $-3.14 \cdot 10^{-4}$

27. _____

28. _____

Concept Connections

29. What is the difference between 4^4 and 4^{-4}?

_____

_____

30. Do $(-x)^2$ and $-x^2$ represent equivalent quantities?
Why or why not?

_____

_____

Chapter 2 VARIABLE SENSE

Activity 2.1

Learning Objectives
1. Identify input and output in situations involving two variable quantities.
2. Determine the replacement values for a variable within a given situation.
3. Use a table to numerically represent a relationship between two variables.
4. Represent a relationship between two variables graphically.
5. Identify trends in data pairs that are represented numerically and graphically.

Key Terms
Use the vocabulary terms listed below to complete each statement in Exercises 1–5.

variable	input	output
horizontal	vertical	ordered pair

1. When viewing a table of paired data values, the _____ is the value that is considered first and the _____ results from the first value.

2. For a plotted point on a rectangular coordinate system, a(n) _____ is called the coordinates of a point.

3. A(n) _____, usually represented by a letter, is a quantity or quality that may change.

4. The output variable is referenced on the _____ axis.

5. The input variable is referenced on the _____ axis.

Practice Exercises

Use the following simulated data to answer #6-18. The Smiley-face Sunglass Company sells fashionable sunglasses. The following graph shows the number of sunglasses sold from 2000 through 2009.

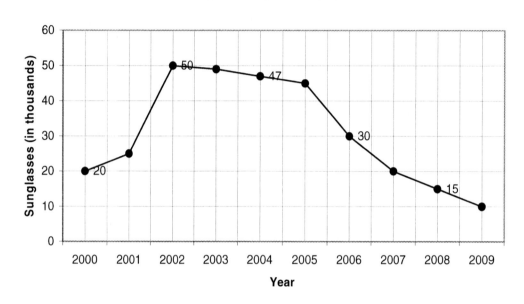

Smiley-face Sunglass Company

6. Identify the output variable.

7. Which axis represents the output variable?

6. _____

7. _____

8. Identify the input variable.

9. Which axis represents the input variable?

8. _____

9. _____

10. Complete the following table, listing the number of sunglasses sold in each given year.

10. _____

Year, y	Sunglasses, s (in thousands)
2001	
2003	
2005	
2007	
2009	

11. What letter in the table from #10 is used to represent the input variable?

12. What letter in the table from #10 is used to represent the output variable?

11. _____

12. _____

13. Estimate the year in which 49,000 sunglasses were sold.

14. Estimate the year in which 10,000 sunglasses were sold.

13. _____

14. _____

15. During what years did sales decrease the least?

16. For which one-year period does the graph indicate the most rapid increase of sunglasses sales?

15. _____

16. _____

17. For which one-year period does the graph indicate the most rapid decrease of sunglasses sales?

18. From the graph, predict the sunglasses sales for 2010.

17. _____

18. _____

Use the following statement to answer #19-21.
Suppose the variable f represents the number of frequent flier miles earned by a businessman.

19. Can f be reasonably replaced by negative value, such as –100? Explain.

_____

_____

20. Can f be reasonably replaced by the number 0? Explain.

_____

_____

21. What is a possible collection of replacement values for the variable *f* in this situation?

_____

_____

Use the following statement to answer #22-26.
Suppose the variable *h* represents the elevation in the state of California.

22. Can *h* be reasonably replaced by an elevation of –100 ft? Explain.

_____

_____

23. Can *h* be reasonably replaced by an elevation of –500 ft? Explain.

_____

_____

24. Can *h* be reasonably replaced by an elevation of 0 ft? Explain.

_____

_____

25. Can *h* be reasonably replaced by an elevation of 7,000 ft? Explain.

_____

_____

26. Can *h* be reasonably replaced by an elevation of 20,000 ft? Explain.

_____

_____

27. What is a possible collection of replacement values for the variable h that would make this situation realistic?

_____

_____

Concept Connections
Use the graph given before Exercise #6 to answer the following questions.

28. What do you predict will happen to the Smiley-face Sunglass Company in years 2011 and 2012? What assumptions are you making in your prediction?

_____

_____

29. Explain why the points are evenly spaced on the horizontal axis.

_____

_____

30. In which years were the number of sunglasses sold 20,000? Must each year have a different number of sunglasses sold? Why or why not?

_____

_____

Chapter 2 VARIABLE SENSE

Activity 2.2

Learning Objectives
1. Construct a graph of data pairs using an appropriately scaled and labeled rectangular coordinate system.
2. Determine the coordinates of a point on a graph.
3. Identify points that lie in a given quadrant or on a given axis.

Key Terms
Use the vocabulary terms listed below to complete each statement in Exercises 1–6.

quadrants	**scaling**	**input**	**output**
origin	**horizontal**	**vertical**	

1. The process of determining and labeling an appropriate distance between tick marks is called _____.

2. The point of intersection of the axes is called the _____ and has coordinates $(0, 0)$.

3. The two perpendicular coordinate axes divide the plane into four _____.

4. In an ordered pair of numbers (x, y) the _____ coordinate, x, is written first, and the _____ coordinate, y, is written second.

5. The _____ variable is referenced on the vertical axis.

6. The _____ variable is referenced on the horizontal axis.

Practice Exercises

7. Plot these points. $(2, 4)$, $(-3, 1)$, $(2, -5)$, $(-4, -1)$, $(0, -2)$, $(2, 0)$, $(0, 3)$, $(-4, 0)$

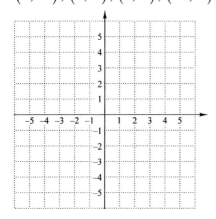

For #8-22, name the quadrant or axis where each point lies.

8. $(-2, 5)$ **9.** $(4, 0)$

8. _____

9. _____

10. $(3, -1)$ **11.** $(6, 6)$

10. _____

11. _____

12. $(0, -2)$ **13.** $(-1, -1)$

12. _____

13. _____

14. $(-5, 0)$ **15.** $(5, 4)$ **14.** _____

                                                           15. _____

16. $(-4, 2)$ **17.** $(-3, -5)$ **16.** _____

                                                           17. _____

18. $(1, -3)$ **19.** $(0, 5)$ **18.** _____

                                                           19. _____

20. $(-1, -3)$ **21.** $(1, -1)$ **20.** _____

                                                           21. _____

22. $(3, 0)$ **22.** _____

For #23-27, find the coordinates of the points A, B, C, D, and E.

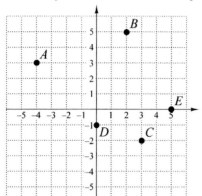

23. A_____

24. B_____

25. C_____

26. D_____

27. E_____

Concept Connections

28. For a point that lies on a horizontal axis, what is the *y*-value? For a point that lies on a vertical axis, what is the *x*-value?

_____

_____

29. Draw a rectangular coordinate system that has proper scaling and plot the points (0, 0), (1, 1), (2, 2), (3, 3), (4, 4) and (5, 5). Next draw another rectangular coordinate system that does NOT have equal spaces between tick marks, and plot the same points. Compare the two graphs.

_____

_____

30. Sue can't remember which value goes first when plotting points. What verbal rule can you state that explains which coordinate to plot first and its location on the graph?

_____

_____

Chapter 2 VARIABLE SENSE

Activity 2.3

Learning Objectives
1. Identify input variables and output variables.
2. Determine possible replacement values for the input.
3. Write verbal rules that represent relationships between input and output variables.
4. Construct tables of input/output variables.
5. Construct graphs from input/output tables.

Key Terms
Use the vocabulary terms listed below to complete each statement in Exercises 1–2.

    verbal   **replacement**

1. A _____ value is an input value for which a meaningful output value can be determined.
2. A _____ rule is a statement that describes the arithmetic steps used to calculate the output corresponding to any input value.

Practice Exercises

3. Determine the arithmetic relationship common to all input and output pairs in the table. Complete the table.

Input	2	3	4	5	6	7
Output	6	9	12			

4. Write the arithmetic relationship common to all input and output pairs in the table.

_____

5. Graph the input/output data from #3 on an appropriately scaled and labeled set of coordinate axes.

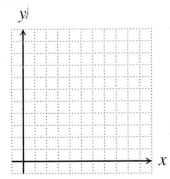

6. Determine the arithmetic relationship common to all input and output pairs in the table. Complete the table.

Input	−1	1	3	5	7	9
Output	4	6	8			

7. Write the arithmetic relationship common to all input and output pairs in the table.

_____

8. Graph the input/output data from #6 on an appropriately scaled and labeled set of coordinate axes.

9. Determine the arithmetic relationship common to all input and output pairs in the table. Complete the table.

Input	−3	0	3	6	9	12
Output	−12	0	12			

10. Write the arithmetic relationship common to all input and output pairs in the table.

_____

11. Determine the arithmetic relationship common to all input
and output pairs in the table. Complete the table.

Input	-3	-1	1	3	5	7
Output	-8	-6	-4			

12. Write the arithmetic relationship common to all input and output pairs in the table.

_____

13. Determine the arithmetic relationship common to all input
and output pairs in the table. Complete the table.

Input	-4	-2	0	2	4	8
Output	2	4	6			

14. Write the arithmetic relationship common to all input and output pairs in the table.

_____

15. Determine the arithmetic relationship common to all input
and output pairs in the table. Complete the table.

Input	-4	-2	0	2	4	6
Output	-2	-1	0			

16. Write the arithmetic relationship common to all input and output pairs in the table.

_____

17. Determine the arithmetic relationship common to all input and output pairs in the table. Complete the table.

Input	−3	−1	1	3	6	7
Output	9	1	1			

18. Write the arithmetic relationship common to all input and output pairs in the table.

---

19. Complete the table.

Input	Output is 3 less than the input
0	
3	
6	
9	
12	

20. Graph the information in the completed table from #19 on an appropriately scaled and labeled set of coordinate axes.

21. Complete the table.

Input	Output is 2 times the input plus 7
−4	
−3	
−2	
−1	
0	
1	

22. Graph the information in the completed table from #21 on an appropriately scaled and labeled set of coordinate axes.

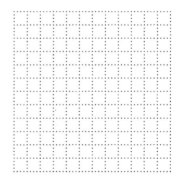

For #23-25, use the following information. Grapes are $1.50 per pound, including tax.

23. How much would you pay for 4 pounds of grapes?   **23.** _____

24. Write a verbal rule that describes the cost for an amount of grapes.

_____

_____

25. Complete the table.

Pounds of grapes	2	3	6	8	10
Cost					

26. A brand new car is valued at $24,000. Each year the value of the car decreases by $2000. Complete the table.

Year	2	4	6	8	10
Value of car					

27. Complete the table.

Input	Output is 5 times the input minus 4
−3	
−2	
−1	
0	
1	
2	

Concept Connections

28. Using the information from #26, is zero a reasonable replacement value for input? Explain.

_____

_____

29. Using the information from #23-25, are negative values reasonable values for input? Explain.

_____

_____

30. Using only the graph from #5, what would be the next input/output values? How did you make that prediction?

_____

_____

Chapter 2 VARIABLE SENSE

Activity 2.4

Learning Objectives
1. Generalize from an arithmetic calculation to a symbolic representation by utilizing variables.
2. Evaluate algebraic expressions.

Key Terms
Use the vocabulary terms listed below to complete each statement in Exercises 1–4.

evaluate	algebraic expression	terms	coefficient

1. A numerical constant that is multiplied by a variable is called the _____ of the variable.

2. A symbolic representation is often called a(n) _____ in the variable x.

3. The _____ of an algebraic expression are the parts of the expression that are added or subtracted.

4. We _____ an algebraic expression when a replacement value is substituted for the variables, and the sequence of arithmetic operations is performed to produce a single output value.

Practice Exercises
For #5–12, let x represent the input variable. Translate each of the phrases into an algebraic expression.

5. The input is increased by 8

6. The sum of four and one-third the input

5. _____

6. _____

7. Twelve less than the input

8. The quotient of the input and –3

7. _____

8. _____

9. Four less than the product of the input and 7

10. Two times the difference between 10 and the input

9. _____

10. _____

11. Nine more than five times the input

12. Thirteen times the input plus 8

11. _____

12. _____

For #13-23, evaluate each algebraic expression for the given values.

13. $4x - 3$ for $x = -2$

14. $6 - 1.2x$ for $x = 5$

13. _____

14. _____

15. $\frac{1}{2}n(n-1)$ for $n = 6$

16. $\frac{1}{3}rh$ for $r = 9$ and $h = 4$

15. _____

16. _____

17. $7x + 2$ for $x = 3$

18. $10y$ for $y = 4$

17. _____

18. _____

19. $2.5 - 3.6x$ for $x = 5$ **20.** $4.7y - 2.1$ for $y = -2$ **19.** _____

                                                                                20. _____

21. $14w - 5$ for $w = 2$ **22.** $4 - 15y$ for $y = 0$ **21.** _____

                                                                                22. _____

23. $7 - 2.2x$ for $x = 5$ **23.** _____

24. A 10% markup of every toy in the toy store is 110% of the
original price of each item. If p represents the original
price, write an algebraic expression that can be used to
determine the markup.

_____

25. Determine the value of a   **26.** Determine the value of a   **25.** _____
$20 toy after a 10% $44 toy after a 10%
markup.   markup.

                                                                                26. _____

27. An isosceles triangle has two equal sides. If one side has length d and the two equal sides each have length $2d$, write an algebraic expression that represents the perimeter of the triangle.

28. If the length of the equal sides of the triangle from #27 is increased by 3, represent the new perimeter in terms of d.

27. _____

28. _____

Concept Connections

You have pennies, nickels, dimes and quarters in a jar.

29. What variable letters would you use, and what algebraic expression should be used to symbolically represent this value?

_____

_____

30. An ice cream cone at a fast food establishment costs $1.20. If you have 8 pennies, 4 nickels, 9 dimes and 2 quarters, do you have enough money to buy the ice cream cone?

_____

_____

Chapter 2 VARIABLE SENSE

Activity 2.6

Learning Objectives
1. Translate verbal rules into symbolic rules.
2. Solve an equation of the form $ax = b$, $a \neq 0$, for x using an algebraic approach.
3. Solve an equation of the form $x + a = b$ for x using an algebraic approach.

Key Terms

Use the vocabulary terms listed below to complete each statement in Exercises 1–6.

solution	**multiply**	**divide**
add	**subtract**	**algebraic approach**

1. To undo addition of a number to the input, _____ that number from each side of the equation.

2. To undo multiplication of the input by a nonzero number, _____ each side of the equation by that number.

3. To undo subtraction of a number from the input, _____ that number to each side of the equation.

4. To undo division of the input by a number, _____ each side of the equation by that number.

5. A(n) _____ of an equation containing one variable is a replacement value for the variable that produces equal values on both sides of the equation.

6. A(n) _____ to solving an equation for a given variable is complete when the variable is isolated on one side of the equation with coefficient 1.

Practice Exercises

In exercises #7-30, for the odd-numbered exercises, determine the output when you are given the input. For the even-numbered exercises, use an algebraic approach to solve the equation for the input when you are given the output.

7. Given $4.5x = y$, determine y when $x = 18$.

8. Given $4.5x = y$, determine x when $y = 126$.

7. _____

8. _____

9. Given $z = -15x$,
    determine z
    when $x = -11$.

10. Given $z = -15x$,
    determine x
    when $z = 105$.

9. _____

10. _____

11. Given $y = 14.6x$,
    determine y
    when $x = -12$.

12. Given $y = 14.6x$,
    determine x
    when $y = 306.6$.

11. _____

12. _____

13. Given $y = x + 9$,
    determine y
    when $x = -17$.

14. Given $y = x + 9$,
    determine x
    when $y = 14$.

13. _____

14. _____

15. Given $y = x + 7.3$,
    determine y
    when $x = -3.9$.

16. Given $y = x + 7.3$,
    determine x
    when $y = 19.7$.

15. _____

16. _____

17. Given $z = x - 21$,
    determine z
    when $x = -7$.

18. Given $z = x - 21$,
    determine x
    when $z = -11$.

17. _____

18. _____

19. Given $9.3x = y$,
determine y
when $x = 23$.

20. Given $9.3x = y$,
determine x
when $y = 502.2$.

19. _____

20. _____

21. Given $z = -13.5x$,
determine z
when $x = -24$.

22. Given $z = -13.5x$,
determine x
when $z = 332.1$.

21. _____

22. _____

23. Given $y = x + 17$,
determine y
when $x = -21$.

24. Given $y = x + 17$,
determine x
when $y = 24$.

23. _____

24. _____

25. Given $y = x + 2.3$,
determine y
when $x = -0.7$.

26. Given $y = x + 2.3$,
determine x
when $y = 3.8$.

25. _____

26. _____

27. Given $y = 13.2x$,
determine y
when $x = -24$.

27. _____

Concept Connections

A family had lunch at the Angry Trout. The cost of the meal was $30.00. They would like to add a tip of 15%.

28. What is the amount of the tip?

_____

_____

29. The entire bill is paid by a credit card with the tip included, what is the total amount charged?

_____

_____

30. Write a verbal rule to determine the total bill charged.

_____

_____

Chapter 2 VARIABLE SENSE

Activity 2.7

Learning Objectives
1. Model contextual situations with symbolic rules of the form $y = ax + b$, $a \neq 0$.
2. Solve equations of the form $ax + b = c$, $a \neq 0$.

Key Terms

Use the vocabulary terms listed below to complete each statement in Exercises 1–3.

inverse	**reverse**	**divide**
add	**subtract**	**multiply**

1. To solve an equation containing two or more operations, perform the sequence of operations in _____ order, replacing each original operation with its

   _____.

2. To solve $p = 3q - 5$ for q, first _____ 5 and

   then _____ by 3.

3. To solve $m = \dfrac{n}{2} + 7$ for n, first _____ 7 and

   then _____ by 2.

Practice Exercises

For #4-13, solve each of the following equations for x.

4. $17 = 2x + 5$ 5. $-37 = -4x - 17$ 4. _____

                                                          5. _____

6. $4x - 13 = 15$ 7. $26 - 3x = 38$ 6. _____

                                                        7. _____

8. $6x - 13 = 26$ **9.** $-3x + 7 = 7$ **8.** _____

9. _____

10. $13 + \dfrac{1}{4}x = 8$ **11.** $\dfrac{3}{4}x - 15 = 0$ **10.** _____

11. _____

12. $0.45x - 1.6 = 2$ **13.** $6 = 3.6x - 30$ **12.** _____

13. _____

For #14-20, replace y by the given value and solve the resulting equation for x.

14. If $y = 4x - 6$ and   **15.** If $y = 20 - 3x$ and   **14.** _____
$y = 18$, determine x. $y = 11$, determine x.

15. _____

16. If $y = \dfrac{2}{3}x - 23$ and   **17.** If $y = 20 - \dfrac{5}{4}x$ and

  $y = -17$, determine x. $y = 5$, determine x.

18. If $2.4x + 7 = y$ and   **19.** If $-4x + 25 = y$ and

  $y = 5$, determine x. $y = -3$, determine x.

20. If $y = 7 - 11x$ and $y = -4$, determine x.

For #21-26, complete the table.

21. $y = 5x - 9$

x	y
7	
	56

22. $y = 15 + 1.5x$

x	y
0.5	
	-15

23. $y = -4x + 25$

x	y
$\dfrac{3}{4}$	
	-7

24. $y = 13 - \dfrac{4}{5}x$

x	y
-10	
	1

25. $y = \dfrac{x}{6} - 3$

x	y
36	
	12

26. $y = 8 - 2.4x$

x	y
5	
	-13

27. $y = 5x - 9$

x	y
2	
	-29

27. _____

Concept Connections

When solving an equation, the sequence of operations is performed in reverse order.

28. Kim wants to solve $y = 3x - 5$ for x. First she divides by 3, and then adds 5. What is her mistake?

_____

29. If the inverse of multiplication is division and the inverse of division is multiplication, what additional restrictions are placed on these operations?

_____

_____

30. Explain how an equation is like a balance when solving an equation.

_____

_____

Chapter 2 VARIABLE SENSE

Activity 2.8

Learning Objectives
1. Evaluate formulas for specified input values.
2. Solve a formula for a specified variable.

Practice Exercises

For #1-3, use the formula $c = 8d$.

1. Solve for d.   **2.** If $d = 5$, find c.   **1.** _____

                                                                    2. _____

3. If $c = 72$, find d.   **3.** _____

For #4-6, use the formula $y = 19 - x$.

4. Solve for x.   **5.** If $x = 14$, find y.   **4.** _____

                                                                    5. _____

6. If $y = 24$, find x.   **6.** _____

7. For the formula $A = \dfrac{p + q + r}{3}$, solve for p.   **7.** _____

For #8-10, use the formula $11a = 12b$.

8. Solve for b.   **9.** If $b = 22$, find a.

8. _____

9. _____

10. If $a = 60$, find b.

10. _____

For #11-13, use the formula $A = 4\pi r^2$.

11. Solve for r^2.   **12.** If $r = 6$, find A as a multiple of π.

11. _____

12. _____

13. If $A = 100\pi$, find r.

13. _____

For #14-16, use the formula $c = \pi d$.

14. Solve for d.   **15.** If $d = 7$, find c as a multiple of π.

14. _____

15. _____

16. If $c = 13\pi$, find d.

16. _____

For #17-19, use the formula $P = 4s$.

17. Solve for s.   **18.** If $s = 9$, find P.   17. _____

18. _____

19. If $P = 48$, find s.   19. _____

For #20-28, solve each formula for the indicated variable.

20. $c = \dfrac{4k}{w}$ for w **21.** $t = \dfrac{xy}{z}$ for z 20. _____

21. _____

22. $m = \dfrac{x - y}{2}$ for x **23.** $P = ax + b$ for x 22. _____

23. _____

24. $ab + ac = d$ for c **25.** $G = w + 150n$ for n 24. _____

25. _____

26. $F = ma$ for a **27.** $V = lwh$ for h **26.** _____

27. _____

28. $P = 2w + 2h + l$ for l **28.** _____

Concept Connections

29. Jill is to solve $y = 2x + 6$ for x when $y = 2$. She knows that there are two different procedures she can use to find the answer: (1) solve the equation for x first and then substitute for y, or (2) substitute for y and then solve for x. Which procedure is more efficient in this case?

_____

_____

30. Jill is to solve $y = 2x + 6$ for x when $y = 1$, $y = 3$, and $y = 5$. She knows that there are two different procedures she can use to find the answer: (1) solve the equation for x first and then substitute for y, or (2) substitute for y and then solve for x. Which procedure is more efficient in this case?

_____

_____

Chapter 2 VARIABLE SENSE

Activity 2.9

Learning Objectives
1. Recognize that equivalent fractions lead to proportions.
2. Use proportions to solve problems involving ratios and rates.
3. Solve proportions.

Practice Exercises

For #1-12, solve each proportion for x.

1. $\dfrac{3}{11} = \dfrac{x}{88}$ **2.** $\dfrac{9}{7} = \dfrac{54}{x}$

1. _____

2. _____

3. $\dfrac{x}{21} = \dfrac{100}{70}$ **4.** $\dfrac{5}{13} = \dfrac{x}{65}$

3. _____

4. _____

5. $\dfrac{7}{18} = \dfrac{21}{x}$ **6.** $\dfrac{153}{x} = \dfrac{17}{15}$

5. _____

6. _____

7. $\dfrac{x}{204} = \dfrac{87}{34}$ **8.** $\dfrac{78}{x} = \dfrac{138}{23}$

9. $\dfrac{120}{x} = \dfrac{3}{40}$ **10.** $\dfrac{360}{x} = \dfrac{9}{5}$

9. _____

10. _____

11. $\dfrac{13}{8} = \dfrac{x}{120}$ **12.** $\dfrac{16}{23} = \dfrac{48}{x}$

11. _____

12. _____

For #13-20, solve.

13. A car travels 102 miles on 3 gallons of gas. How far   **13.** _____
can it travel on 11 gallons of gas?

14. Three gallons of paint cover 1350 square feet of wall.   **14.** _____
How many gallons of paint are needed for 3150 square
feet of wall?

15. Dan drove his new car 17,500 miles in the first 5 months.   **15.** _____
At this rate, how many miles will he drive in 2 years?

16. The scale for a map is 1.5 inches equals 50 miles. If two   **16.** _____
cities are 6 inches apart on the map, find the distance
between them.

17. A stock split of 5 shares for every 2 shares owned. If Greg   **17.** _____
owned 350 shares, how many shares will he have after the
split?

18. In a pizza dough recipe, 1.5 teaspoons of salt are mixed   **18.** _____
with 4 cups of flour. How much salt is needed for 20 cups
of flour?

19. A Little League team won $\frac{3}{5}$ of the games played. They   **19.** _____

played 175 games. How many games did they win?

20. In an online math class 16 students earned an A, which   **20.** _____

was $\frac{4}{9}$ of the class. How many students were enrolled in

the class?

For #21-28, determine whether each proportion is true or false.

21. $\dfrac{27}{8} = \dfrac{18}{5}$ **22.** $\dfrac{225}{675} = \dfrac{3}{9}$

21. _____

22. _____

23. $\dfrac{39}{54} = \dfrac{52}{72}$ **24.** $\dfrac{7}{4} = \dfrac{42}{20}$

23. _____

24. _____

25. $\dfrac{450}{7} = \dfrac{300}{4}$ **26.** $\dfrac{28}{12} = \dfrac{7}{3}$

25. _____

26. _____

27. $\dfrac{304}{13} = \dfrac{47}{2}$ **28.** $\dfrac{57}{3} = \dfrac{133}{7}$

27. _____

28. _____

Name:

Date:

Instructor:

Section:

Concept Connections

29. Bill says that $\dfrac{a}{b} = \dfrac{c}{d}$ is equivalent to $\dfrac{d}{c} = \dfrac{b}{a}$. Matt disagrees. Who is right? Why?

_____

_____

30. Jody says that $\dfrac{a}{b} = \dfrac{c}{d}$ is equivalent to $\dfrac{a}{d} = \dfrac{b}{c}$. Barbara disagrees. Who is right? Why?

_____

_____

Chapter 2 VARIABLE SENSE

Activity 2.10

Learning Objectives
1. Translate verbal rules into symbolic (algebraic) rules.
2. Write algebraic expressions that involve grouping symbols.
3. Evaluate algebraic expressions containing two or more operations.
4. Identify equivalent algebraic expressions by examining their outputs.

Practice Exercises

For #1-10, let x represent the input variable and y represent the output variable.
Translate each of the verbal rules into a symbolic rule.

1. The output is the input decreased by nine.

2. The output is five times the difference between the input and seven.

1. _____

2. _____

3. The output is eight increased by the quotient and the input and three.

4. The output is six more than one-third of the square of the input.

3. _____

4. _____

5. The output is sixteen less than the product of the input and –5.

6. The output is the sum of one-fourth of the input and fourteen.

5. _____

6. _____

7. The output is –8 times the sum of the input and eleven.

8. The output is thirteen decreased by the quotient of the input and twelve.

9. The output is seventeen less than the product of the input and six.

10. The output is eight less than twice the square of the input.

11. Complete the table.

x	$x \cdot 4$	$4x$
-3		
-1		
0		
2		

12. Do the expressions $x \cdot 4$ and $4x$ produce the same output value when given the same input value?

13. What property of multiplication is demonstrated in Exercise #12?

14. Use a graphing calculator to sketch $y_1 = x \cdot 4$ and $y_2 = 4x$ on the same coordinate axes. How do the graphs compare?

15. Is the expression $x - 5$ equivalent to $5 - x$? Complete the table.

x	$x - 5$	$5 - x$
-6		
-1		
0		
1		
6		

16. From Exercise 15, what correspondence do you observe between the output values in columns 2 and 3? How is the expression related to the expression $x - 5$?

16. _____

17. Is the operation of subtraction commutative?

18. Use your graphing calculator to sketch $y_1 = x - 5$ and $y_2 = 5 - x$ on the same coordinate axes. How do the graphs compare?

17. _____

18. _____

19.

x	$y_1 = 4x - 1$	$y_2 = 4(x - 1)$
-2		
0		
2		
3		

20. Diagram the rule in Exercise 19.

20. _____

21. Are the expressions $4x-1$ and $4(x-1)$ equivalent?
Why or why not?

_____

_____

22. Complete the table.

x	$y_3 = x^2 - 1$	$y_4 = (x-1)^2$
-1		
0		
2		
4		

23. Diagram the rule in Exercise 22.   23. _____

24. Are the expressions $x^2 - 1$ and $(x-1)^2$ equivalent?
Why or why not?

_____

_____

25. Complete the table.

x	$y_5 = \dfrac{1}{2}x^2 + 1$	$y_6 = \dfrac{1}{2}(x^2 + 2)$
-2		
0		
4		
6		

26. Diagram the rule in Exercise 25.   26. _____

27. Are the expressions $\frac{1}{2}x^2 + 1$ and $\frac{1}{2}(x^2 + 2)$ equivalent?

Why or why not?

_____

_____

Concept Connections

28. You want to multiply $31 \cdot 5$ mentally. How can you
use an equivalent expression to find this product?
Hint: $31 = 30 + 1$.

_____

_____

29. You want to multiply $49 \cdot 3$ mentally. How can you
use an equivalent expression to find this product?
Hint: $49 = 50 - 1$.

_____

_____

30. Joe diagrams the rule for $y = 7(x^2 - 2)$ as

y : start with $x \rightarrow$ square $\rightarrow$ multiply by $7 \rightarrow$ subtract $2 \rightarrow$ to obtain y.

What is Joe doing wrong?

_____

_____

Chapter 2 VARIABLE SENSE

Activity 2.11

Learning Objectives
1. Apply the distributive property.
2. Use areas of rectangles to interpret the distributive property geometrically.
3. Identify equivalent expressions.
4. Identify the greatest common factor in an expression.
5. Factor out the greatest common factor in an expression.
6. Recognize like terms.
7. Simplify an expression by combining like terms.

Key Terms
Use the vocabulary terms listed below to complete each statement in Exercises 1–4.

coefficient	factor	term	common

1. A _____ is either a number or a product of a number and one or more variables.

2. A numerical factor that multiplies a variable term is called the _____ of the variable.

3. A factor that is in each term of an algebraic expression is called a _____ factor.

4. When two or more mathematical expressions are multiplied to form a product, each of the original expressions is called a _____ of the product.

Practice Exercises
For #5-12, use the distributive property to expand each algebraic expression.

5. $5(6x - 2)$ 6. $-4(t - 3.2)$ 5. _____

                                                  6. _____

7. $3.5(7 - 3x)$ 8. $-(8p - 13)$ 7. _____

                                                  8. _____

9. $-(9x-11y-z)$ **10.** $-4(3x^2-5x+6)$ **9.** _____

10. _____

11. $-\dfrac{1}{3}\left(\dfrac{9}{13}x-\dfrac{3}{4}\right)$ **12.** $\dfrac{6}{5}\left(\dfrac{3}{2}x-\dfrac{4}{3}\right)$ **11.** _____

12. _____

For #13-16, identify the greatest common factor and rewrite the expression in completely factored form.

13. $15x-12y$ **14.** $15x^2+20x-5$ **13.** _____

14. _____

15. $21x^2-35x$ **16.** $15a^2-10ab-35a$ **15.** _____

16. _____

For #17-28, use the distributive property, and then combine like terms to simplify the expressions.

17. $15-(x+5)$ **18.** $5x-4(10-x)$ **17.** _____

18. _____

Name:

Instructor:

Date:

Section:

19. $36-4(6x+7y)$

20. $13.6-(4.6-x)$

19. _____

20. _____

21. $7x+3(x+8)$

22. $5(x-3)+6(x+2)$

21. _____

22. _____

23. $3x^2-4x(x-5)$

24. $y(3x-4)-4y(x+5)$

23. _____

24. _____

25. $12(0.4x-2)-(0.8x+3)$

26. $9(x-5)-3(x-7)$

25. _____

26. _____

27. $x+5(3x-11)$

28. $20(1.5x+3)-(10x+15)$

27. _____

28. _____

Concept Connections

29. Which of $5y^2$, $3y^2$, and $3y$ are like terms? Why?

_____

_____

30. After factoring, how can you check your answer?
After using the distributive property, how can you check your answer?

_____

_____

Chapter 2 VARIABLE SENSE

Activity 2.12

Learning Objectives
1. Recognize an algebraic expression as a code of instruction.
2. Simplify algebraic expressions.

Practice Exercises

For #1-28, simplify each expression.

1. $5(3-n+2)-7$ **2.** $6x+3(7x+8)$ **1.** _____

2. _____

3. $4(x+y)+5(2x+7)-5y+6$ **3.** _____

4. $-(x-3y)+4(-2x+5y)+8x$ **4.** _____

5. $5+2[4x+3(x-4)-8(x+2)-6x]$ **5.** _____

6. $7[4+3(x-9)]-[4-(x+6)]$

6. _____

7. $\dfrac{6(x-3)-4x+2}{2}$

7. _____

8. $3\{4-[5(x-6)-2]+7x\}$

8. _____

9. $\dfrac{5(x-7)-2x+8}{3}$

9. _____

10. $[6-3(x-4)]\div3+11$

10. _____

11. $[5-3(x+1)]-[-5(3x-2)+7]+10$

11. _____

12. $\dfrac{5(x-3)+10}{5}+4$

12. _____

13. $\dfrac{(6x-7)+x}{7}+8$

13. _____

14. $\left[(5n+9)-2n\right]\div 3-6$

14. _____

15. $7\left[\dfrac{-3n+9}{3}-4\right]+8$

15. _____

16. $4(-n+3-8)+9$

16. _____

17. $7+8(3x-2)$

17. _____

18. $5(x-y)+6(3x+2)+4y-10$ **18.** _____

19. $\dfrac{4(x-5)+12}{2}+3$ **19.** _____

20. $-(x-5y)+4(-4x+3y)+8x$ **20.** _____

21. $\dfrac{7(x-3)-2x+1}{5}$ **21.** _____

22. $3[6+4(x-2)]-[8-(x+5)]$ **22.** _____

23. $\dfrac{9x+15}{3}-11$ **23.** _____

24. $9x + 3(5x - 7)$

25. $8[2 + 3(x - 4)] - [5 - (x + 2)]$

26. $\dfrac{10(x - 3) - 6x + 2}{4}$

27. $8\{5 - [6(x - 3) - 4] + 7x\}$

28. $\dfrac{2(x + 5) - 2}{2}$

Concept Connections

29. Name some real world examples of instructions where the order of the directions matters.

_____

_____

30. The word *simplify* means many different things. Name some of the operations used when you are asked to simplify an algebraic expression.

_____

_____

Chapter 2 VARIABLE SENSE

Activity 2.13

Learning Objectives

1. Translate verbal rules into symbolic rules.
2. Write and solve equations of the form $ax + b = cx + d$.
3. Use the distributive property to solve equations involving grouping symbols.
4. Develop mathematical models to solve problems.
5. Solve formulas for a specified variable.

Practice Exercises

For #1-16, solve each equation.

1. $7x - 6 = 4x + 12$ **1.** _____

2. $5x - 11 = 7x + 23$ **2.** _____

3. $0.4x + 8 = 5.4x - 12$ **3.** _____

4. $5x - 12 = -3x + 4$ **4.** _____

5. $0.4x - 6.5 = 0.3x + 3.2$

5. _____

6. $5 - 0.03x = 0.2 - 0.05x$

6. _____

7. $8t + 6 = 5(t + 3)$

7. _____

8. $5(w - 3) = w - 11$

8. _____

9. $31 + 4(x - 7) = 5(x + 6)$

9. _____

10. $-81 = -(x - 7) - 15x$

10. _____

11. $17 + 3(5x - 2) = 7x - 13$ **11.** _____

12. $19 + 3(6x - 7) = 18x - 2$ **12.** _____

13. $8 - 7(x - 5) = 12 - 7x$ **13.** _____

14. $3(x + 2) = 4(2x + 1) + 7$ **14.** _____

15. $-57 = -7(x - 3) - 1$ **15.** _____

16. $93 = -(x + 19) + 23$ **16.** _____

For #17-28, solve each formula for the given variable.

17. $y = mx + b$, for m

17. _____

18. $A = \dfrac{B + C}{2}$, for C

18. _____

19. $4x - 3y = 6$, for y

19. _____

20. $C = \dfrac{5}{9}(F - 32)$, for F

20. _____

21. $A = P + Prt$, for r

21. _____

22. $13 = -x + \dfrac{y}{4}$, for y

22. _____

23. $P = 2b + 2s$, for s 23. _____

24. $R = 175 - 0.25a$, for a 24. _____

25. $5x - 6y = 30$, for x 25. _____

26. $7x + 3y = 21$, for x 26. _____

27. $9x - \dfrac{1}{2}y = 10$, for y 27. _____

28. $9x - \dfrac{1}{2}y = 10$, for x 28. _____

Concept Connections

29. When Joan solves the equation $3(2x+3)+2=2(x+4)$, the result is $8 = 8$. What does this mean? What should Joan write as the answer?

_____

_____

30. When Evan solves the equation $-3x+4=-1-3x$, the result is $4 = -1$. What does this mean? What should Evan write as the answer?

_____

_____

Name: Date:
Instructor: Section:

Chapter 3 FUNCTION SENSE AND LINEAR FUNCTIONS

Activity 3.1

Learning Objectives
1. Describe in words what a graphs tells you about a given situation.
2. Sketch a graph that best represents a situation that is described in words.
3. Identify increasing, decreasing, and constant parts of a graph.
4. Identify minimum and maximum points on a graph.
5. Define a function.
6. Use the vertical line test to determine whether a graph represents a function.

Practice Exercises
For #1-20, use the vertical line test to determine if the graph represents a function.

1.

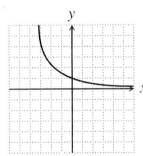

2.

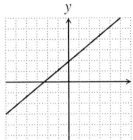

1. _____

2. _____

3.

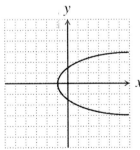

4.

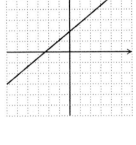

3. _____

4. _____

5.

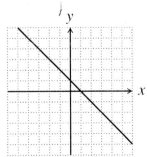

6.

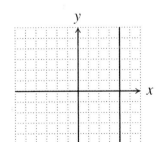

5. _____

6. _____

7.

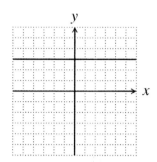

8.

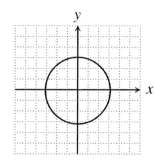

9.

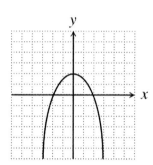

10.

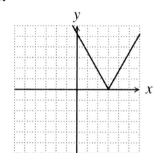

11.

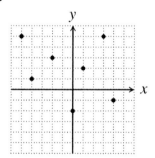

12.

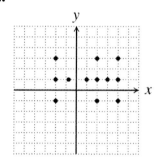

13.

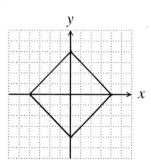

14.

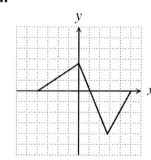

7. _____

8. _____

9. _____

10. _____

11. _____

12. _____

13. _____

14. _____

15.

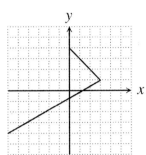

16.

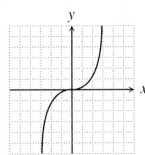

15. _____

16. _____

17.

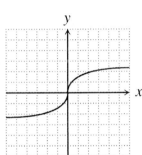

18.

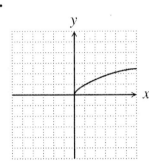

17. _____

18. _____

19.

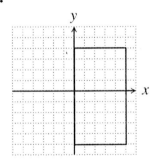

20.

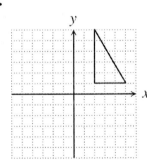

19. _____

20. _____

For #21-28, label each graph as increasing, decreasing, or constant.

21.

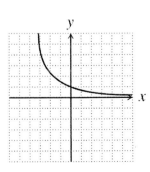

22.

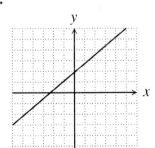

21. _____

22. _____

23.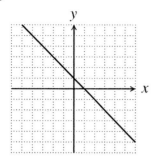

24.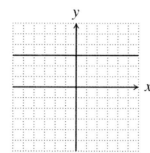

25.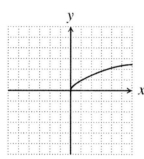

26.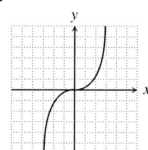

27.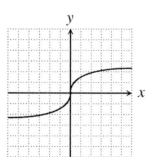

28.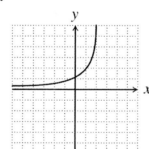

23. _____

24. _____

25. _____

26. _____

27. _____

28. _____

Concept Connections
29. Define function.

_____

_____

30. Under what circumstances is there a local maximum point on a graph?
Under what circumstances is there a local minimum point on a graph?

_____

_____

Chapter 3 FUNCTION SENSE AND LINEAR FUNCTIONS

Activity 3.2

Learning Objectives
1. Represent functions numerically, graphically, and symbolically.
2. Determine the symbolic rule that defines a function.
3. Use function notation to represent functions symbolically.
4. Identify the domain and range of a function.
5. Identify the practical domain and range of a function.

Practice Exercises

For #1-10, evaluate each function.

1. Let $f(x) = 3x - 1$.

Evaluate $f(5)$.

2. Let $g(n) = 6n + 4$.

Evaluate $g(-2)$.

1. _____

2. _____

3. Let
$h(m) = 3m^2 - 2m + 6$.
Evaluate $h(-3)$.

4. Let $p(x) = 4x^2 + 3x + 7$.
Evaluate $p(-2)$.

3. _____

4. _____

5. Let $f(a) = 7a + 2$.

Evaluate $f(-4)$.

6. Let $g(b) = 11b - 10$.

Evaluate $g(2)$.

5. _____

6. _____

7. Let $h(c) = 4c^2 - 7c + 2$.   **8.** Let $p(x) = 6x^2 + 8x - 5$.   **7.** _____
Evaluate $h(3)$.   Evaluate $p(-5)$.

8. _____

9. Let $g(x) = -4x + 9$.   **10.** Let $f(x) = \dfrac{x}{5}$.   **9.** _____
Evaluate $g(3)$.   Evaluate $f(-10)$.

10. _____

11. Let f be a function defined by $f(x) = 3x - 1$.   **11.** _____
Determine the value of x for which $f(x) = 5$.

12. Let g be a function defined by $g(x) = 7x - 5$.   **12.** _____
Determine the value of x for which $g(x) = -26$.

13. Let h be a function defined by $h(t) = \dfrac{t}{6}$.   **13.** _____
Determine the value of t for which $h(t) = 12$.

14. Let k be a function defined by $k(w) = 0.7w$.
Determine the value of w for which $k(w) = 56$.

14. _____

15. Let H be a function defined by $H(a) = -5a + 8$.
Determine the value of a for which $H(a) = 23$.

15. _____

16. Let f be a function defined by $f(x) = -5x + 2$.
Determine the value of x for which $f(x) = 22$.

16. _____

17. Let g be a function defined by $g(x) = \frac{1}{2}x - 3$.
Determine the value of x for which $g(x) = 1$.

17. _____

18. Let h be a function defined by $h(t) = \frac{t}{10}$.
Determine the value of t for which $h(t) = 45$.

18. _____

19. Let k be a function defined by $k(w) = 1.4w$.
Determine the value of w for which $k(w) = 21$.

19. _____

20. Let H be a function defined by $H(a) = 10a - 6$.
Determine the value of a for which $H(a) = -36$.

20. _____

For #21-28, consider the function defined by the following table and assume that these points are the only input/output pairs that belong to the function f.

x	-4	-2	0	2	4	6
$f(x)$	6	3	-2	1	5	7

21. What is the domain of f? **22.** What is the range of f?

21. _____

22. _____

23. For which set of consecutive x-values is f increasing? **24.** Determine the maximum output of f and the input for which it occurs.

23. _____

24. _____

25. Determine the minimum output value of f and the input for which it occurs. **26.** Write f as a set of ordered pairs.

25. _____

26. _____

27. Determine $f(-4)$. **28.** For what value of x is $f(x) = -2$.

27. _____

28. _____

Concept Connections

29. Explain the difference between the domain and the practical domain.

_____

_____

_____

_____

30. Give an example of a situation where the range is different from the practical range.

_____

_____

Chapter 3 FUNCTION SENSE AND LINEAR FUNCTIONS

Activity 3.3

Learning Objectives
1. Determine the average rate of change of an output variable with respect to the input variable.

Key Terms
Use the vocabulary terms listed below to complete each statement in Exercises 1–3.

increases   **decreases**   **remains constant**

1. A line segment connecting two points _____ from left to right if $\dfrac{\Delta y}{\Delta x} = 0$.

2. A line segment connecting two points _____ from left to right if $\dfrac{\Delta y}{\Delta x} > 0$.

3. A line segment connecting two points _____ from left to right if $\dfrac{\Delta y}{\Delta x} < 0$.

Practice Exercises
For #4-7, use the points given below.

x	4	6
y	8	9

4. Find Δx.

5. Find Δy.

4. _____

5. _____

6. Find $\dfrac{\Delta y}{\Delta x}$.

7. Is $\dfrac{\Delta y}{\Delta x}$ from #6 increasing, decreasing or constant?

6. _____

7. _____

For #8-11, use the points given below.

x	1	5
y	6	2

8. Find Δx.

9. Find Δy.

8. _____

9. _____

10. Find $\dfrac{\Delta y}{\Delta x}$.

11. Is $\dfrac{\Delta y}{\Delta x}$ from #10 increasing, decreasing or constant?

10. _____

11. _____

For #12-15, use the points given below.

t	13	18
w	10	20

12. Find Δt.

13. Find Δw.

12. _____

13. _____

14. Find $\dfrac{\Delta w}{\Delta t}$.

15. Is $\dfrac{\Delta w}{\Delta t}$ from #14 increasing, decreasing or constant?

14. _____

15. _____

For #16-19, use the points given below.

x	-1	9
y	9	9

16. Find Δx.

17. Find Δy.

16. _____

17. _____

18. Find $\dfrac{\Delta y}{\Delta x}$.

19. Is $\dfrac{\Delta y}{\Delta x}$ from #18 increasing, decreasing or constant?

18. _____

19. _____

For #20-28, find $\dfrac{\Delta y}{\Delta x}$ from the given points.

20. $(1, 5)$ and $(2, 8)$

21. $(3, 7)$ and $(0, 5)$

20. _____

21. _____

22. $(110, -14)$ and $(6, -2)$

23. $(-4, -2)$ and $(6, 9)$

22. _____

23. _____

24. $(0, 0)$ and $(-1, 5)$

25. $(26, 15)$ and $(15, 81)$

24. _____

25. _____

26. (212, 32) and (100, 0)   **27.** (9, 8) and (0, 0)

26. _____

27. _____

28. (−4, 2) and (−3, 13)

28. _____

Concept Connections

29. Define the average rate of change with respect to input and output.

_____

_____

30. Explain what the symbol Δ is and how it is used in this Activity.

_____

_____

Chapter 3 FUNCTION SENSE AND LINEAR FUNCTIONS

Activity 3.4

Learning Objectives
1. Identify linear functions by a constant average rate of change of the output variable with respect to the input variable.
2. Determine the slope of the line drawn through two points.
3. Identify increasing linear functions using slope.

Practice Exercises

For #1–6, calculate the average rate of change between consecutive data points to determine whether the output in each table is a linear function of the input.

1.

Input	−7	0	7	21
Output	14	28	42	70

1. _____

2.

Input	−3	0	3	9
Output	−30	−5	20	70

2. _____

3.

Input	3	8	13	28
Output	0	5	10	25

3. _____

4. | Input | 1 | 7 | 11 | 17 |
|---|---|---|---|---|
| Output | 2 | 6 | 8 | 10 |

4. _____

5. | Input | −4 | −2 | 1 | 5 |
|---|---|---|---|---|
| Output | −29.6 | −14.8 | 7.4 | 37 |

5. _____

6. | Input | −2 | 0 | 1 | 2 |
|---|---|---|---|---|
| Output | $-\dfrac{1}{2}$ | $\dfrac{3}{2}$ | $\dfrac{7}{4}$ | $\dfrac{9}{4}$ |

6. _____

For #7-12, given a point and slope, determine two additional points on the line. Answers may vary.

7. point: (−6, 11)

slope: $\dfrac{3}{4}$

8. point: (2, −3)

slope: $\dfrac{4}{5}$

7. _____

8. _____

Name:

Date:

Instructor:

Section:

9. point: (3, 2)
slope: 4

10. point: (1, 5)
slope: $\dfrac{1}{4}$

9. _____

10. _____

11. point: (0, 6)
slope: $\dfrac{2}{5}$

12. point: (−3, −1)
slope: 2

11. _____

12. _____

13. Sketch the line from Exercise #9.

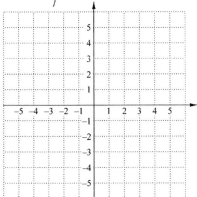

14. Sketch the line from Exercise #12.

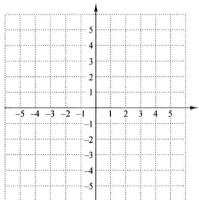

For #15-24, find the slope of the given two points.

15. (2, –3) and (4, 7)   **16.** (–4, 1) and (5, 8)   **15.** _____

                                                                      16. _____

17. (6, –7) and (8, –2)   **18.** (5, 1) and (12, 3)   **17.** _____

                                                                      18. _____

19. (0, 6) and (−2, 0)   **20.** (1.5, 4.2) and (3.5, 7.2)   **19.** _____

                                                                              20. _____

21. (−2, −3) and (2, 1)   **22.** (3, 2) and (7, 8)   **21.** _____

                                                                              22. _____

23. (1, −1) and (9, 5)   **24.** (−4, −4) and (0, 8)   **23.** _____

                                                                              24. _____

25. What is the grade of a road that rises 25 feet over a   **25.** _____
distance of 400 feet?

26. What is the grade of a road that rises 36 feet over a
distance of 450 feet?

26. _____

27. What is the grade of a ramp that rises 2 inches over a
distance of 40 inches?

27. _____

28. What is the grade of a ramp that rises 12 inches over a
distance of 150 inches?

28. _____

Concept Connections

29. Write the symbolic definition for the slope of any two points on the line.

_____

_____

30. How can you use the slope of a line and a point on the line to find another point on the
same line?

_____

_____

Chapter 3 FUNCTION SENSE AND LINEAR FUNCTIONS

Activity 3.5

Learning Objectives
1. Identify lines as having negative, zero, or undefined slopes.
2. Identify a decreasing linear function from its graph or slope.
3. Determine horizontal and vertical intercepts of a linear function from its graph.
4. Interpret the meaning of horizontal and vertical intercepts of a line.

Key Terms
Use the vocabulary terms listed below to complete each statement in Exercises 1–4.

horizontal   **vertical**   **first**   **second**

1. The slope of a _____ line is undefined.

2. The slope of a _____ line is zero.

3. For any vertical intercept, the _____ coordinate is zero.

4. For any horizontal intercept, the _____ coordinate is zero.

Practice Exercises
For #5-14, find the slope of each graph.

5.   **6.**   **5.** _____

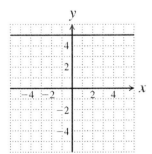   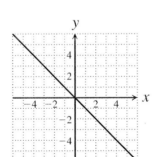

                                                                    6. _____

7.   **8.**   **7.** _____

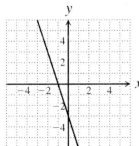   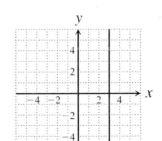

                                                                    8. _____

9.

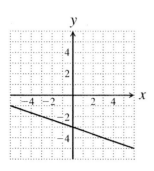

10.

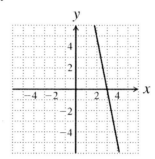

11.

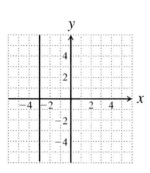

12.

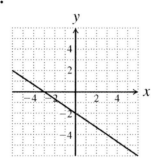

13.

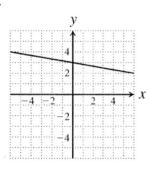

14.

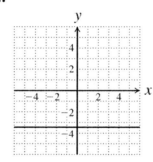

For #15-19, find the horizontal intercept of each graph.

15.

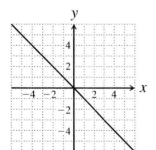

16.

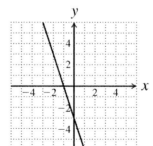

9. _____

10. _____

11. _____

12. _____

13. _____

14. _____

15. _____

16. _____

Name:

Instructor:

Date:

Section:

17.

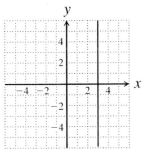

18.

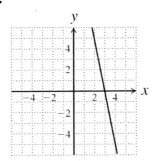

17. _____

18. _____

19.

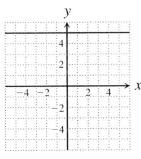

19. _____

For #20-24, find the vertical intercept of each graph.

20.

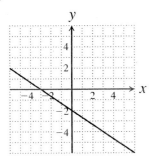

21.

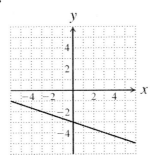

20. _____

21. _____

22.

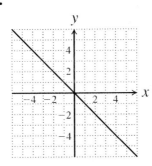

23.

22. _____

23. _____

24.

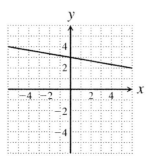

24. _____

For #25-28, answers may vary.

25. A horizontal line contains the point (–2, 6). Determine and list three additional points on the line.

25. _____

26. A line contains the point (3, –2) and has a slope of $-\dfrac{5}{4}$. Determine two additional points on the line.

26. _____

27. A line contains the point (–1, 1) and has a slope of $-\dfrac{2}{3}$. Determine two additional points on the line.

27. _____

28. A vertical line contains the point (1, –5). Determine and list three additional points on the line.

28. _____

Name: Date:
Instructor: Section:

Concept Connections

29. Water freezes at 0°C. One day in January, the outside temperatures decreased (remarkably) as a linear function. At 5 PM it was 10°C. At 7 PM, it was 2°C. At 9 PM, it was –6°C and at 11 PM, it was –14°C. About what time did the water on the street start to freeze? Give an answer to the nearest half hour. (Hint: graph the data.)

_____

_____

30. From Exercise #29, what values did you graph as the horizontal and vertical intercepts?

_____

_____

Chapter 3 FUNCTION SENSE AND LINEAR FUNCTIONS

Activity 3.6

Learning Objectives
1. Determine a symbolic rule for a linear function from contextual information.
2. Identify the practical meanings of the slope and intercepts of a linear function.
3. Determine the slope-intercept form of a linear function.
4. Identify functions as linear by numerical, graphical, and algebraic characteristics.

Practice Exercises

For #1-10, for each linear function, identify the slope and vertical intercept.

1. $y = 5x + 2$ **2.** $y = -3x + 9$ **1.** _____

                                                                                2. _____

3. $y = \dfrac{1}{3}x - 8$ **4.** $y = -8x$ **3.** _____

                                                                                4. _____

5. $y = 25$ **6.** $y = 18x - 96$ **5.** _____

                                                                                6. _____

7. $y = 210 - 7x$ **8.** $p = -13 + 7.6n$ **7.** _____

                                                                                8. _____

9. $q = 64 - 8r$ **10.** $y = 4$ **9.** _____

10. _____

For #11-20, determine an equation of the line from the given slope and y-intercept.

11. Slope: 3
    y-intercept: (0, –2)

12. Slope: –4
    y-intercept: (0, 6)

11. _____

12. _____

13. Slope: $\dfrac{4}{5}$
    y-intercept: (0, 7)

14. Slope: 0
    y-intercept: (0, –9)

13. _____

14. _____

15. Slope: $-\dfrac{2}{5}$
    y-intercept: (0, 0)

16. Slope: 1
    y-intercept: (0, 0)

15. _____

16. _____

17. Slope: 0
    y-intercept: (0, 0)

18. Slope: 0.7
    y-intercept: (0, 7)

17. _____

18. _____

19. Slope: –5
    y-intercept: (0, –5)

20. Slope: 3
    y-intercept: (0, –2)

19. _____

20. _____

For #21-28, determine if the given point lies on the given line.

21. $(-1, 5)$

$y = -2x + 3$

22. $(2, 7)$

$y = -4x + 1$

21. _____

22. _____

23. $(2, 5)$

$y = 6x - 7$

24. $(3, 0)$

$y = -2x + 6$

23. _____

24. _____

25. $(2, 2)$

$y = -x$

26. $(-3, 2)$

$y = -4x - 10$

25. _____

26. _____

27. $(1, 3)$

$y = 5x - 1$

28. $(-1, 5)$

$y = 2x + 7$

27. _____

28. _____

Concept Connections

29. Write the slope-intercept form of the general equation of a line.

_____

30. Given the linear function, $y = x - 5$, name two points on the line. Show that the slope between these two points is the same as the slope of the line.

_____

_____

Chapter 3 FUNCTION SENSE AND LINEAR FUNCTIONS

Activity 3.7

Learning Objectives
1. Identify the slope and vertical intercept from the equation of a line written in slope-intercept form.
2. Write an equation of a line in slope-intercept form.
3. Use the y-intercept and the slope to graph a linear function.
4. Determine horizontal intercepts of linear functions using an algebraic approach.
5. Use intercepts to graph a linear function.

Practice Exercises

For #1-8, find the slope, y-intercept and x-intercept of the given line.

1. $y = 4x - 3$ **2.** $y = -2x + 5$ **1.** _____

                                                        2. _____

3. $y = -x + 3$ **4.** $y = -3$ **3.** _____

                                                        4. _____

5. $y = -\dfrac{x}{2} + 1$ **6.** $3x - 2y = 6$ **5.** _____

                                                        6. _____

7. $4x + 3y = 12$ **8.** $2x - y = 10$

For #9-12, write the equation of the line in slope-intercept form.

9. $2x + y = 9$ **10.** $9x = 4 - 3y$

9. _____

10. _____

11. $5x - 2y = 10$ **12.** $9 - 3y = 0$

11. _____

12. _____

13. What is the slope of the line that goes through the points $(0, 7)$ and $(2, 13)$?

13. _____

14. What is the equation of the line that goes through the points from Exercise #13?

14. _____

15. What is the slope of the line that goes through the points $(-5, 0)$ and $(0, -5)$?

15. _____

16. What is the equation of the line that goes through the points from Exercise #15?

16. _____

17. What is the slope of the line that goes through the points
(0, 7) and (5, 7)?

17. _____

18. What is the equation of the line that goes through the
points from Exercise #17?

18. _____

19. What is the slope of the line that goes through the points
(4, 0) and (4, −10)?

19. _____

20. What is the equation of the line that goes through the
points from Exercise #19?

20. _____

For #21-28, graph each line using the given information.

21. Slope: −3
   y-intercept: (0, −2)

22. Slope: 0
   y-intercept: (0, −1)

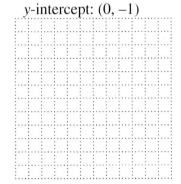

23. Slope: −5
   y-intercept: (0, 0)

24. Slope: 1
   y-intercept: (0, −3)

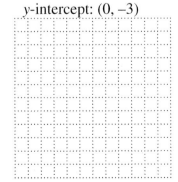

25. x-intercept: (3, 0)
    y-intercept: (0, −4)

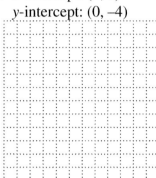

26. x-intercept: (−4, 0)
    y-intercept: (0, 5)

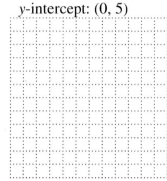

27. x-intercept: (−2, 0)
    y-intercept: (0, −5)

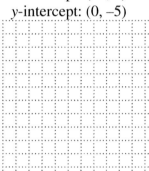

28. x-intercept: (2, 0)
    y-intercept: (0, 2)

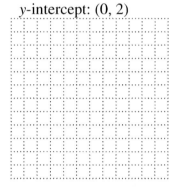

Concept Connections

29. What is the advantage of writing a line in slope-intercept form?

_____

_____

30. Given the horizontal and vertical intercepts $(a, 0)$ and $(0, b)$, find the slope of the line that contains these two points?

_____

_____

Chapter 3 FUNCTION SENSE AND LINEAR FUNCTIONS

Activity 3.8

Learning Objectives
1. Write an equation for a linear function given its slope and y-intercept.
2. Write linear functions in slope-intercept form, $y = mx + b$.
3. Interpret the slope and y-intercept of linear functions in contextual situations.
4. Use the slope-intercept form of linear equations to solve problems.

Practice Exercises

1. Two points on a line are (0, 5) and (8, 21). Determine the slope and y-intercept.

   1. _____

2. Write the equation of the line from Exercise #1 in slope-intercept form.

   2. _____

3. Two points on a line are (0, –6) and (4, 4). Determine the slope and y-intercept.

   3. _____

4. Write the equation of the line from Exercise #3 in slope-intercept form.

   4. _____

5. Two points on a line are (0, 7) and (7, 0). Determine the slope and y-intercept.

   5. _____

6. Write the equation of the line from Exercise #5 in slope-intercept form.

   6. _____

7. Two points on a line are (0, 8) and (10, 8). Determine the slope and y-intercept.

7. _____

8. Write the equation of the line from Exercise #7 in slope-intercept form.

8. _____

9. Two points on a line are $\left(0, \dfrac{1}{2}\right)$ and $\left(5, \dfrac{3}{2}\right)$. Determine the slope and y-intercept.

9. _____

10. Write the equation of the line from Exercise #9 in slope-intercept form.

10. _____

11. Two points on a line are (0, 2) and (–7, 9). Determine the slope and y-intercept.

11. _____

12. Write the equation of the line from Exercise #11 in slope-intercept form.

12. _____

13. Two points on a line are (0, 3) and (4, 3). Determine the slope and y-intercept.

13. _____

14. Write the equation of the line from Exercise #13 in slope-intercept form.

14. _____

15. Two points on a line are (0, 0) and (4, 4). Determine the slope and y-intercept.

15. _____

16. Write the equation of the line from Exercise #15 in slope-intercept form.

16. _____

17. Two points on a line are (0, −2) and (3, 7). Determine the slope and *y*-intercept.

17. _____

18. Write the equation of the line from Exercise #17 in slope-intercept form.

18. _____

19. Two points on a line are (0, 9) and (−3, 3). Determine the slope and *y*-intercept.

19. _____

20. Write the equation of the line from Exercise #19 in slope-intercept form.

20. _____

21. Two points on a line are (0, −5) and (−6, −2). Determine the slope and *y*-intercept.

21. _____

22. Write the equation of the line from Exercise #21 in slope-intercept form.

22. _____

23. Two points on a line are (0, 4) and (10, −8). Determine the slope and *y*-intercept.

23. _____

24. Write the equation of the line from Exercise #23 in slope-intercept form.

24. _____

25. Two points on a line are (0, 1) and (−1, 4). Determine the slope and *y*-intercept.

25. _____

26. Write the equation of the line from Exercise #25 in slope-intercept form.

26. _____

27. Two points on a line are $(0, -3)$ and $(3, 2)$. Determine the slope and y-intercept.

27. _____

28. Write the equation of the line from Exercise #27 in slope-intercept form.

28. _____

Concept Connections

29. When finding the slope of the line connecting the points $(4, -3)$ and $(6, -1)$, Elaine chooses $(x_2, y_2) = (6, -1)$ and $(x_1, y_1) = (4, -3)$. Joanne says that that is one way to find the slope, but insists that she could also choose $(x_2, y_2) = (4, -3)$ and $(x_1, y_1) = (6, -1)$ to get the same answer. Show that Joanne is correct.

_____

_____

30. Jeff insists that the formula for calculating slope is $m = \dfrac{x_2 - x_1}{y_2 - y_1}$, which is incorrect. If Jeff starts with the y-intercept $(0, 0)$ and another point $(1, 5)$, calculates the slope and uses only the incorrect slope and y-intercept to draw his graph, what can be said about his incorrect graph? (Hint: graph the line connecting the two points and then use Jeff's method to graph a second line.)

_____

_____

Chapter 3 FUNCTION SENSE AND LINEAR FUNCTIONS

Activity 3.9

Learning Objectives
1. Determine the slope and y-intercept of a line algebraically and graphically.
2. Determine the equation for a linear function when given two points.
3. Interpret the slope and y-intercept of a linear function in contextual situations.

Practice Exercises

1. Determine the equation of the line with slope 2 that passes through $(3, 1)$.

1. _____

2. Determine the equation of the line with slope -3 that passes through $(4, 1)$.

2. _____

3. Determine the equation of the line with slope $\dfrac{1}{3}$ that passes through $(3, -2)$.

3. _____

4. Determine the equation of the line with slope 0.4 that passes through $(5, 2)$.

4. _____

5. Determine the equation of the line with slope 0 that passes through $(3, 5)$.

5. _____

6. Determine the equation of the line with slope -5.2 that passes through $(2, -5.4)$.

6. _____

7. Determine the equation of the line with slope 0 that passes through (0.5, 6).

7. _____

8. Determine the equation of the line with slope $-\dfrac{4}{3}$ that passes through (−3, 9).

8. _____

For #9-16, use the graph to find the slope and y-intercept of each line.

9.

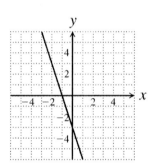

10.

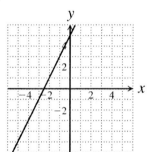

9. _____

10. _____

11.

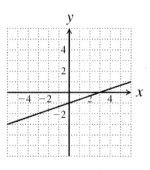

12.

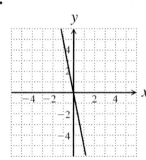

11. _____

12. _____

13.

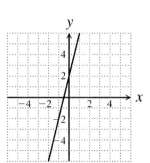

14.

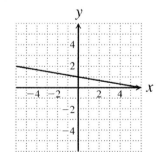

13. _____

14. _____

Name:

Date:

Instructor:

Section:

15.

16.

15. _____

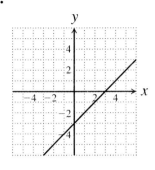

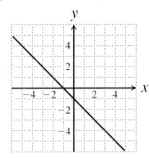

16. _____

For #17-28, determine the equation of the line that passes through the given points.

17. (5, 2) and (10, 6)   **18.** (0, 7) and (−8, 0)   **17.** _____

18. _____

19. (11, 6) and (−7, 6)   **20.** (4, 7) and (9, 17)   **19.** _____

20. _____

21. (−6, 13) and (6, −13)   **22.** (5.5, 7.4) and (4, 2.6)   **21.** _____

22. _____

23. (4, −4) and (6, −4)   **24.** (25, 13) and (−25, 8)   **23.** _____

24. _____

25. (5, 7) and (12, –7) **26.** (4, 2) and (–6, –8) **25.** _____

26. _____

27. (–1, 5) and (2, –4) **28.** (5, 3) and (–5, 0) **27.** _____

28. _____

Concept Connections

29. Write the formula for point-slope form of the equation of a line.

_____

_____

30. Under what circumstances is it easier to use the point-slope equation of a line rather than the slope-intercept equation of a line?

_____

_____

Chapter 3 FUNCTION SENSE AND LINEAR FUNCTIONS

Activity 3.12

Learning Objectives
1. Solve a system of two linear equations numerically.
2. Solve a system of two linear equations graphically.
3. Solve a system of two linear equations symbolically by the substitution method.
4. Recognize the connections among three methods of solution.
5. Interpret the solution to a system of two linear equations in terms of the problem's content.

Key Terms

Use the vocabulary terms listed below to complete each statement in Exercises 1–2.

    consistent   **inconsistent**

1. A linear system is _____ if there is at least one solution, the point of intersection of the graphs.

2. A linear system is _____ if there is no solution. In this case, the lines are parallel.

Practice Exercises

3. Solve $y = 5 - x$
$\qquad y = x + 1$

4. Solve $x = y + 6$
$\qquad x = 12 - 2y$

3. _____

4. _____

5. Check #3 by graphing.

6. Check #4 by graphing.

7. Solve $y = 3x - 3$
$\qquad y = 5x + 3$

8. Solve $q = p + 5$
$\qquad q = 2p - 1$

7. _____

8. _____

9. Check #7 by graphing.

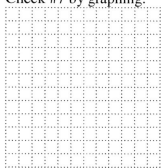

10. Check #8 by graphing.

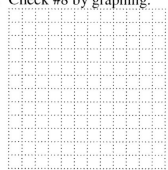

11. Solve $y = -2 - 3.5x$
$\qquad y = 8x - 2$

12. Solve $a = -b - 6$
$\qquad a = b - 8$

11. _____

12. _____

13. Check #11 by graphing.

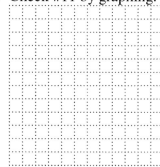

14. Check #12 by graphing.

15. Solve $x = 3y + 4$

           $x = 3y - 5$

16. Solve $x = \frac{1}{2}y + 3$

           $x = 3y - 2$

15. _____

16. _____

17. Check #15 by graphing.

18. Check #16 by graphing.

19. Solve $y = 1 - 4x$

           $y = 2x - 2$

20. Solve $x = 2y$

           $x = 2y + 2$

19. _____

20. _____

21. Check #19 by graphing.

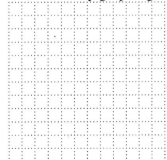

22. Check #20 by graphing.

23. Solve $y = 6x - 1$
$y = 8 - 3x$

24. Solve $x = 2y$
$x = 3y + 1$

23. _____

24. _____

25. Check #23 by graphing.

26. Check #24 by graphing.

27. Solve $y = -2x + 5$
$y = x - 1$

28. Check #27 by graphing.

27. _____

Concept Connections

29. After solving a system of equations, Tom checks his answer only in the first equation, but not in the second equation. What risk is Tom taking?

_____

_____

30. Name a disadvantage to solving a system of equations by graphing.

_____

_____

Chapter 3 FUNCTION SENSE AND LINEAR FUNCTIONS

Activity 3.13

Learning Objectives
1. Solve a system of two linear equations algebraically using the substitution method.
2. Solve a system of two linear equations algebraically using the addition (or elimination) method.

Practice Exercises

For #1-10, solve the system of equations using the substitution method.

1. $y = 3x - 2$
    $2y - x = 1$

1. _____

2. $x = 3y + 7$
    $3x - 5y = 1$

2. _____

3. $7x + 2y = -4$
    $y = 8x - 2$

3. _____

4. $x + y = 1$
$x = 4 - 2y$

4. _____

5. $x + y = 4$
$y = x + 1$

5. _____

6. $x = y - 3$
$x + 3y = 9$

6. _____

7. $3x + 4 = x$
$3y - x = 5$

7. _____

8. $y = 3x + 3$
         $2x - y = 6$

8. _____

9. $x = 2y - 7.4$
         $y - 5x = 10$

9. _____

10. $x - 6 = -2y$
         $3x - 3y = 18$

10. _____

For #11-20, solve the system of equations using the elimination method.

11. $x - y = 3$
         $x + y = 11$

11. _____

12. $3x - y = 2$
$-5x + y = 4$

12.   _____

13. $x - 3y = 7$
$6x - 10y = 2$

13.   _____

14. $3x - 2y = 5$
$6x - 9 = 4y$

14.   _____

15. $x + y = 8$
$-x + 5y = 7$

15.   _____

16. $x + y = 9$
         $3x - y = -5$

16. _____

17. $-x + y = 6$
         $x + 2y = 6$

17. _____

18. $x + 5y = 2$
         $2x - 3y = 17$

18. _____

19. $2x - 3y = 3$
         $3x + 3y = 12$

19. _____

20. $4x + 5y = 6$

 $\quad x + y = 0$

20. _____

For #21-28, solve the system of equations using either substitution or elimination.

21. $\quad y + 8 = x$

 $\quad 5x + y = -2$

21. _____

22. $\quad 2x - y = 5$

 $\quad 3x - 4y = -5$

22. _____

23. $x + 5y = 1$

 $\quad x + y = -3$

23. _____

24. $5x + y = 11$
$x + 3y = 33$

24. _____

25. $x - 2y = 2$
$4x - 5y = 14$

25. _____

26. $8x + 12y = 10$
$6x + 9y = 5$

26. _____

27. $y - x = 2$
$2x + 3y = -4$

27. _____

28. $3x - 2y = 0.4$
$5x \; - \; y = 9.3$

28. _____

Concept Connections

29. Greg says, just by looking at it, that he can tell the system

$y = 5x - 1$
$y = 5x + 3$

has no solution. How did he come to that conclusion?

_____

_____

30. Which method do you prefer: substitution or elimination?

_____

_____

Chapter 3 FUNCTION SENSE AND LINEAR FUNCTIONS

Activity 3.15

Learning Objectives
1.   Use properties of inequalities to solve linear inequalities in one variable graphically.

Practice Exercises

For #1-28, solve each inequality.

1. $x - 8 > 17$

2. $2x + 7 \leq 3x + 2$

1. _____

2. _____

3. $-7 + x \geq 14$

4. $-5x > -13$

3. _____

4. _____

5. $x - \dfrac{1}{5} < \dfrac{1}{10}$

6. $-\dfrac{2}{3} \geq 12x$

5. _____

6. _____

7. $-4x < 7$

8. $2 + 3x > 17$

7. _____

8. _____

9. $11x - 18 < -51$ **10.** $7x + 8 - 6x \geq 12$ **9.** _____

10. _____

11. $12 - 10x \leq 3 - 9x$ **12.** $13 - 8y < y + 40$ **11.** _____

12. _____

13. $\dfrac{x}{4} - 6 > 2$ **14.** $4(3y - 5) \leq 4$ **13.** _____

14. _____

15. $2(x - 4) + 5 \geq 7(x + 8) - 19$ **15.** _____

16. $12 - (3x + 4) < 5(x + 7) + x$ **16.** _____

17. $13x - 2 - 7x \geq 2(3x - 1) + 5 + x$ **17.** _____

18. $3(x+2) < 8(x+7)$ **18.** _____

19. $7(4x+6) \le 10(3x+2)$ **19.** _____

20. $8-5(2x-3) \ge 23-2x$ **20.** _____

21. $2[5-3(4-x)]-7 < 3[2(5x-1)+8]-15$ **21.** _____

22. $\dfrac{x}{3}+1 \le 4x+12$ **22.** _____

    175

23. $9 - x \geq 3(9 - x)$

24. $5.5 - 2x \leq 0.5x + 3$

23. _____

24. _____

25. $6 - x < 3(x - 4) + 2$

26. $3.2 - 3x \leq 0.4x + 6.6$

25. _____

26. _____

27. $3(x - 4) > 2(x - 6)$

28. $1 - 2.4x > 9.6x - 5$

27. _____

28. _____

Concept Connections
29. Explain the difference between $x > 2$ and $x \geq 2$.

_____

30. Under what circumstances is the direction of an inequality reversed?

_____

_____

Chapter 4 AN INTRODUCTION TO NONLINEAR PROBLEM SOLVING

Activity 4.1

Learning Objectives
1. Identify polynomials and polynomial functions.
2. Classify a polynomial as a monomial, binomial, or trinomial.
3. Determine the degree of a polynomial.
4. Simplify a polynomial by identifying and combining like terms.
5. Add and subtract polynomials.
6. Evaluate and interpret polynomials.

Key Terms
Use the vocabulary terms listed below to complete each statement in Exercises 1–5.

degree	**standard form**	**monomial**
binomial	**trinomial**	

1. The _____ of a monomial is defined as the exponent on its variable.

2. A _____ is a polynomial with three terms.

3. A _____ is a polynomial with one term.

4. A _____ is a polynomial with two terms.

5. _____ is when all the terms of a polynomial are written with the powers of x in decreasing order.

Practice Exercises
For #6-9, perform the indicated operations, and express your answer in simplest form.

6. $7x + 3(4x - 5)$ 7. $4.2(a + b) + 3.9b$ 6. _____

7. _____

8. $6.7 - (4.7 - x)$ 9. $5(0.4x + 2) - 4(0.3x + 7)$ 8. _____

9. _____

For #10-14, evaluate the function at the given value.

10. $f(x) = 4x^2 - 5x + 3$

Determine $f(3)$.

11. $S(d) = 0.6d^2 + 5d$

Determine $S(2.5)$.

10. _____

11. _____

12. $C(x) = 5x^3 - 16$

Determine $C(6)$.

13. $p(a) = 4(a - 2.3)$

Determine $p(3.7)$.

12. _____

13. _____

14. $D(n) = 7(n - 15)$

Determine $D(-8)$.

14. _____

For #15-19, add the following polynomials.

15. $9x^3 + 13x - 6$ and
$-5x^2 - 7x + 15$

16. $-7x^2 + 3x - 5$ and
$6x^2 - 9x + 8$

15. _____

16. _____

17. $35x^3 - 15x^2 - 19$ and
$-40x^3 + 25x^2 - 11x + 19$

18. $3x^2 + 4x - 14$ and
$-8x^2 + 5x - 9$ and
$6x^2 - 4x + 12$

17. _____

18. _____

19. $6x^2 - 2x + 11$ and $-11x^2 + 4x - 15$ and $3x^2 + 2x + 5$ **19.** _____

For #20-24, perform the indicated operations. Express your answer in simplest form.

20. $\left(17x^2 + 13x - 7\right) - \left(8x^2 - 4x + 3\right)$ **20.** _____

21. $\left(5x^3 + x^2 - 11\right) - \left(4x^2 - 13x + 2\right)$ **21.** _____

22. $\left(7x^2 - 9x + 5\right) - \left(8x^2 + 2x - 4\right) - \left(2x^2 - 5x + 9\right)$ **22.** _____

23. $\left(-4x^3 - 6x^2 - 9x - 7\right) - \left(-4x^3 - 6x^2 - 9x + 7\right)$ **23.** _____

24. $\left(11x^2 + 7x - 8\right) - \left(3x^2 - 3x - 5\right) + \left(8x^2 - 10x + 3\right)$ **24.** _____

For #25-28, determine the degree of each polynomial. Write the polynomial in standard form.

25. $15 - 7x^3 + 19x$ **26.** $-8x + 17x^2 - 11$

25. _____

26. _____

27. $14 + 4x^6 - 3x + 19x^3$ **28.** -11

27. _____

28. _____

Concept Connections

29. Bill says that $\sqrt{3} + x$ and $\sqrt{3 + x}$ are both polynomials. Paul says only one is a polynomial. Who is right, and why?

_____

_____

30. The function for profit is $P(x) = R(x) - C(x)$, where x is the number of items made. The Major Widget company produces 3500 mini-widgets with a profit of \$80,500. Write the function using these values.

_____

_____

Chapter 4 AN INTRODUCTION TO NONLINEAR PROBLEM SOLVING

Activity 4.2

Learning Objectives
1. Use properties of exponents to simplify expressions and combine powers that have the same base.
2. Use the distributive property and properties of exponents to write expressions in expanded form.

Practice Exercises

For #1-18, use the properties of exponents to simplify each expression.

1. $4xx^5$

2. $5t^6 7t^3$

3. $-8w^3 6w^7$

4. $2.5b^6 1.02b^5$

5. $6(x^3)^5$

6. $(-x^{15})^3$

7. $(-6x^4)(0.4x^7)(3.5x^5)$

8. $(a^4 b^2 c^5)(a^3 c^7)$

1. _____

2. _____

3. _____

4. _____

5. _____

6. _____

7. _____

8. _____

9. $(-3s^3t^2)(t^3)^2(s^5t)$ **10.** $\dfrac{15x^5y^7}{3x^4y^6}$ **9.** _____

10. _____

11. $\dfrac{6w^3z^4}{15w^5z^4}$ **12.** $\dfrac{x^5}{x^9}$ **11.** _____

12. _____

13. $\left(\dfrac{5a^3}{3b^2}\right)^3$ **14.** $6a\left(3a^4\right)^2$ **13.** _____

14. _____

15. $12x^3\left(y^5\right)^0$ **16.** $\dfrac{x^0}{x^5}$ **15.** _____

16. _____

17. $\dfrac{x^{11}}{x^{15}}$ **18.** $\dfrac{14w^7}{2w^0}$ **17.** _____

18. _____

Name: Date:

Instructor: Section:

For #19-28, use the distributive property and the properties of exponents to expand each algebraic expression and write it as a polynomial in standard form.

19. $5x(x+7)$ **20.** $y(4y-1)$ **19.** _____

                                                     20. _____

21. $x^3(3x^2+4x-1)$ **22.** $r^5(7.4r-2.3)$ **21.** _____

                                                       22. _____

23. $4t^3(5t^5-3t^3-1.2)$ **24.** $2.4x^9(-3x^4-5x^2+1)$ **23.** _____

                                                       24. _____

25. $7x^4(3x-14)$ **26.** $3a(a^5+5a^3-6)$ **25.** _____

                                                       26. _____

27. $5x^2(9x^3-4x^2-2.4)$ **28.** $y^0(4y^7-3y^2+9)$ **27.** _____

                                                       28. _____

    183

Concept Connections

29. Jeff has a hard time remembering the rules for exponents. How would you help him remember how to simplify the expressions $\left(x^3\right)^4$ and $y^5 y^3$?

_____

_____

30. For y^5, identify the base and exponent. Write y^5 as a product of two or more quantities.

_____

_____

Chapter 4 AN INTRODUCTION TO NONLINEAR PROBLEM SOLVING

Activity 4.3

Learning Objectives
1. Expand and simplify the product of two binomials.
2. Expand and simplify the product of any two polynomials.
3. Recognize and expand the product of conjugate binomials: difference of squares.
4. Recognize and expand the product of identical binomials: perfect-square trinomials.

Key Terms
Use the vocabulary terms listed below to complete each statement in Exercises 1–2.

   difference of squares   **perfect-square trinomial**

1. The product $(x + y)(x - y)$ results in a _____ .

2. The product $(x + y)(x + y)$ results in a _____ .

Practice Exercises
For #3-14, multiply each of the following pairs of binomial expressions. Remember to simplify by combining like terms.

3. $(x + 2)(x + 8)$ **4.** $(w - 6)(w - 4)$ **3.** _____

     4. _____

5. $(x - 5)(x + 8)$ **6.** $(6 + 5c)(3 + c)$ **5.** _____

     6. _____

7. $(x + 4)(x - 4)$ **8.** $(7x - 1)(3x + 4)$ **7.** _____

     8. _____

9. $(4x+3)(3x+1)$ **10.** $(5a-4)(7a+2)$ **9.** _____

10. _____

11. $(4x-3)(x-4)$ **12.** $(5c+2d)(5c-2d)$ **11.** _____

12. _____

13. $(3a-2b)(2a-3b)$ **14.** $(x+5y)(x-4y)$ **13.** _____

14. _____

For #15-20, multiply each expression. Write your answer in simplest form.

15. $(x+3)(x^2-4x+6)$ **16.** $(x-5)(4x^2-x+3)$ **15.** _____

16. _____

17. $(x-5y)(2x-xy-5y)$ **18.** $(2a+3b)(a^2-3b-1)$ **17.** _____

18. _____

19. $(x-4)(x^2+4x+16)$ **20.** $(x+5)(x^2-5x+25)$

19. _____

20. _____

For #21-28, expand and simplify the following binomial products.

21. $(x-5)^2$ **22.** $(x+6)^2$

21. _____

22. _____

23. $(5x-4)(5x+4)$ **24.** $(6x+7)(6x-7)$

23. _____

24. _____

25. $(x-25)^2$ **26.** $(x+13)^2$

25. _____

26. _____

27. $(x-11)^2$ **28.** $(3x-8)(3x+8)$

27. _____

28. _____

Concept Connections

29. Define conjugate binomials.

_____

_____

30. FOIL is a term commonly used when multiplying two binomials. What does FOIL stand for?

_____

_____

Name: Date:
Instructor: Section:

Chapter 4 AN INTRODUCTION TO NONLINEAR PROBLEM SOLVING

Activity 4.4

Learning Objectives
1. Evaluate quadratic functions of the form $y = ax^2$. Note: $a \neq 0$.
2. Graph quadratic functions of the form $y = ax^2$. Note: $a \neq 0$.
3. Interpret the coordinates of points on the graph of $y = ax^2$ in context. Note: $a \neq 0$.
4. Solve a quadratic equation of the form $ax^2 = c$ graphically. Note: $a \neq 0$.
5. Solve a quadratic equation of the form $ax^2 = c$ algebraically by taking square roots. Note: $a \neq 0$.
6. Solve a quadratic equation of the form $(x \pm a)^2 = c$ algebraically by taking square roots.

Practice Exercises
For #1-5, solve each equation. Write decimal results to the nearest hundredth.

1. $6x^2 = 96$ **2.** $-36x^2 = 180$ **1.** _____

                                                          2. _____

3. $\dfrac{x^2}{3} = 12$ **4.** $3x^2 = 300$ **3.** _____

                                                          4. _____

5. $4x^2 = 60$ **5.** _____

For #6-10, solve each equation by first writing the equation in the form $x^2 = c$.

6. $5x^2 + 7 = 7$ **7.** $16 + c^2 = 65$ **6.** _____

7. _____

8. $4a^2 - 51 = 93$ **9.** $t^2 - 64 = 0$ **8.** _____

9. _____

10. $5a^2 + 74 = 94$ **10.** _____

For #11-27, solve each equation.

11. $(x + 3)^2 = 25$ **12.** $(x - 1)^2 = 49$ **11.** _____

12. _____

13. $(x - 5)^2 = 100$ **14.** $(x - 6)^2 = 36$ **13.** _____

14. _____

15. $6(x - 2)^2 = 96$ **16.** $(5 + x)^2 + 30 = 151$ **15.** _____

16. _____

17. $(3x-4)^2 = 64$

18. $(5x-3)^2 = 49$

17. _____

18. _____

19. $(4x+11)^2 = 121$

20. $(6x-5)^2 = 100$

19. _____

20. _____

21. $3(7x-2)^2 = 243$

22. $4(9x-5)^2 = 196$

21. _____

22. _____

23. $(9+x)^2 - 30 = 70$

24. $2(3x-5)^2 + 41 = 91$

23. _____

24. _____

25. $(x-4)^2 = 225$

26. $(2x-1)^2 = 81$

25. _____

26. _____

27. $(x+3)^2 = 144$

27. _____

Concept Connections

28. What does $\sqrt{y}$ represent?

_____

_____

29. Quadratic functions have a specific shape and solution. Name that shape and explain what the solutions mean graphically.

_____

_____

30. Explain what the symbol $\pm$ means, and give an example of a solution using $\pm$.

_____

_____

Chapter 4 AN INTRODUCTION TO NONLINEAR PROBLEM SOLVING

Activity 4.5

Learning Objectives

1. Evaluate functions of the form $y = ax^2 + bx$, $a \neq 0$.

2. Graph functions of the form $y = ax^2 + bx$, $a \neq 0$.

3. Identify the x-intercepts of the graph of $y = ax^2 + bx$ graphically and algebraically.

4. Interpret the x-intercepts of a quadratic function in context.

5. Factor a binomials of the form $ax^2 + bx$.

6. Solve an equation of the form $ax^2 + bx = 0$ using the zero-product property.

Practice Exercises

For #1-6, use the zero-product property to solve each equation.

1. $x(x+15) = 0$ **2.** $2x(x-11) = 0$

1. _____

2. _____

3. $(x-4)(x+5) = 0$ **4.** $(3x+2)(x-6) = 0$

3. _____

4. _____

5. $(9x-1)(2x+5) = 0$ **6.** $(4x-3)(x+1) = 0$

5. _____

6. _____

For #7-14, use the greatest common factor to rewrite each equation in factored form, and then apply the zero-product property to find the solutions.

7. $x^2 - 5x = 0$ **8.** $x^2 + 7x = 0$ **7.** _____

                                                      8. _____

9. $2x^2 - 14x = 0$ **10.** $6x^2 - 3x = 0$ **9.** _____

                                                    10. _____

11. $x^2 = 11x$ **12.** $4x^2 = 32x$ **11.** _____

                                                    12. _____

13. $8x^2 - 10x = 0$ **14.** $10x^2 = 15x$ **13.** _____

                                                    14. _____

For #15-27, use the zero-product property to determine the solutions of each equation.

15. $x^2 + 25x = 0$ **16.** $6x^2 = 84x$ **15.** _____

                                                    16. _____

Name: Date:

Instructor: Section:

17. $16y^2 + 96y = 0$ **18.** $36t^2 - 54t = 0$ **17.** _____

18. _____

19. $9(x-6)(x+2) = 0$ **20.** $13x = x^2$ **19.** _____

20. _____

21. $11p = 4p - 7p^2$ **22.** $7w^2 - 6w = 8w$ **21.** _____

22. _____

23. $23p = 4p - 19p^2$ **24.** $4(3x-7)(2x+5) = 0$ **23.** _____

24. _____

25. $5(4x-1)(3x+1) = 0$ **26.** $16y^2 - 48y = 0$ **25.** _____

26. _____

27. $x^2 - 18x = 0$ **27.** _____

Concept Connections

28. When describing a quadratic equation or function, why is it always stated that for $y = ax^2$, $a \neq 0$?

_____

_____

29. State the zero-product property.

_____

_____

30. Explain how to determine the x-intercepts algebraically from $y = ax^2 + bx$.

_____

_____

Chapter 4 AN INTRODUCTION TO NONLINEAR PROBLEM SOLVING

Activity 4.6

Learning Objectives

1. Recognize and write a quadratic equation in standard form, $ax^2 + bx + c = 0$, $a \neq 0$.
2. Factor trinomials of the form $x^2 + bx + c$.
3. Solve a factorable quadratic equation of the form $x^2 + bx + c = 0$ using the zero-product property.
4. Identify a quadratic function from its algebraic form.

Practice Exercises

For #1-28, solve each quadratic equation by factoring.

1. $x^2 + 8x + 7 = 0$ **2.** $x^2 - 11x + 24 = 0$ **1.** _____

                                                                        2. _____

3. $x^2 - x - 20 = 0$ **4.** $x^2 + 17x - 60 = 0$ **3.** _____

                                                                        4. _____

5. $x^2 - 13x + 42 = 0$ **6.** $x^2 + 5 = 2x + 20$ **5.** _____

                                                                        6. _____

7. $y^2 + 10y = -21$

8. $8x^2 + 120x + 288 = 0$

7. _____

8. _____

9. $2x^2 - 18x + 36 = 0$

10. $x^2 - 11x + 30 = 0$

9. _____

10. _____

11. $4x^2 - 32 = 8x$

12. $x^2 + 31 = 32x$

11. _____

12. _____

13. $x^2 - 10x = 27 - 4x$

14. $3x^2 - 9x = 2x^2 + 22$

13. _____

14. _____

15. $-2x^2 + 4x = -70$

16. $3x^2 = 45x - 78$

15. _____

16. _____

17. $x^2 + 10x = 39$

18. $2x^2 - 2x = 84$

17. _____

18. _____

19. $3x^2 = 39x - 120$

20. $x^2 + 40 = -18x - 5$

19. _____

20. _____

21. $-2x^2 + 28x = 96$

22. $3x^2 = 9x + 162$

21. _____

22. _____

23. $-5x^2 + 25x = -30$

24. $4x^2 - 24x = 28$

23. _____

24. _____

25. $2x^2 + 26 = 28x$

26. $3x^2 + 1 = 6x - 2$

25. _____

26. _____

27. $x = 12 - x^2$ **28.** $5x^2 - 40 - 35x = 0$ **27.** _____

28. _____

Concept Connections

29. Write the general form of a quadratic function.

_____

_____

30. Define standard form for a quadratic equation.

_____

_____

Chapter 4 AN INTRODUCTION TO NONLINEAR PROBLEM SOLVING

Activity 4.7

Learning Objectives
1. Use the quadratic formula to solve quadratic equations.
2. Identify the solutions of a quadratic equation with points on the corresponding graph.

Practice Exercises

For #1-28, solve each quadratic equation with the quadratic formula.

1. $3x^2 - 4x + 1 = 0$ **2.** $x^2 - 3x - 18 = 0$

1. _____

2. _____

3. $4x^2 + 40x + 51 = 0$ **4.** $2x^2 + x - 15 = 0$

3. _____

4. _____

5. $4x^2 - 11x = 0$ **6.** $10x^2 - 21x - 49 = 0$

5. _____

6. _____

7. $5x^2 + 9x = 0$

8. $5x^2 + 14x - 55 = 0$

7. _____

8. _____

9. $4x^2 - 15x + 9 = 0$

10. $8x^2 + 20x + 5 = 0$

9. _____

10. _____

11. $x^2 - 9x - 8 = 0$

12. $4x^2 - 6x - 1 = 0$

11. _____

12. _____

13. $x^2 - 6x + 4 = 0$

14. $x^2 + 6x - 13 = 0$

13. _____

14. _____

15. $2x^2 - 15x + 16 = 0$

16. $2x^2 - 3x - 3 = 0$

15. _____

16. _____

17. $7x^2 - 3x - 2 = 0$ **18.** $3x^2 - 8x - 5 = 0$ **17.** _____

                                                                        18. _____

19. $x^2 + 2x - 30 = 0$ **20.** $x^2 - 25x + 100 = 0$ **19.** _____

                                                                        20. _____

21. $x^2 + 24x + 135 = 0$ **22.** $x^2 - 33x + 200 = 0$ **21.** _____

                                                                        22. _____

23. $x^2 + 30x - 136 = 0$ **24.** $35x^2 - 57x - 44 = 0$ **23.** _____

                                                                        24. _____

25. $6x^2 + 19x + 15 = 0$ **26.** $2x^2 - x - 1 = 0$ **25.** _____

                                                                        26. _____

27. $3x^2 + 2x - 5 = 0$ **28.** $2x^2 - x - 15 = 0$ **27.** _____

28. _____

Concept Connections

29. Write the quadratic formula.

_____

_____

30. What does the $\pm$ mean in the quadratic formula?

_____

_____

Chapter 4 AN INTRODUCTION TO NONLINEAR PROBLEM SOLVING

Activity 4.8

Learning Objectives
1. Recognize an exponential function as a rule for applying a growth factor or a decay factor.
2. Graph exponential functions from numerical data.
3. Recognize exponential functions from symbolic rules.
4. Graph exponential functions from symbolic rules.

Practice Exercises

For #1–4, suppose a population of bugs is increasing by 7% per day, and the rate remains the same for the next 5 days.

1. Determine the growth factor.   **1.** _____

2. The population today is 50,000 bugs. If the population   **2.** _____
growth remains constant at 7% per day, write an algebraic
rule to determine the bug population after *t* days.

3. Determine the bug population after 1 day and after 2 days.   **3.** _____

4. Determine the bug population after 4 days and after 5   **4.** _____
days. Round to the nearest bug.

205

For #5–8, some say the number of texts sent will increase 40% each year for the next 5 years.

5. Determine the growth factor.   5. _____

6. The number of texts sent this year could be 230 million. If   6. _____
   the number of texts remains constant at 40% per year,
   write an algebraic rule to determine the number of texts, in
   millions, sent after t years.

7. Determine the number of texts sent after 1 year and after 2   7. _____
   years. Round to the nearest million.

8. Determine the number of texts sent after 3 years and after   8. _____
   4 years. Round to the nearest million.

For #9–12, after purchasing a laptop, you discover that it will depreciate 20% per year for the next five years.

9. Determine the decay factor.   9. _____

10. You purchase the laptop for $1000. If the rate of   10. _____
    depreciation remains constant at 20% per year, write an
    algebraic rule to determine the value of the laptop after t
    years.

11. Determine the value of the laptop after 1 year and after 2 years.

11. _____

12. Determine the value of the laptop after 3 years and after 4 years. Round to the nearest dollar.

12. _____

For #13–16, after using an insecticide, a population of bugs is decreasing by 35% per hour, and the rate remains the same for the next 5 hours.

13. Determine the decay factor.

13. _____

14. The population now is 30,000 bugs. If the population decay remains constant at 35% per hour, write an algebraic rule to determine the bug population after t hours.

14. _____

15. Determine the bug population after 1 hour and after 2 hours.

15. _____

16. Determine the bug population after 4 hours and after 5 hours. Round to the nearest bug.

16. _____

For #17–20, suppose a population of a rapidly expanding city is increasing by 10% per year, and the rate remains the same for the next 5 years.

17. Determine the growth factor.

17. _____

18. The population today is 150,000. If the population growth remains constant at 10% per year, write an algebraic rule to determine the bug population after t years.

18. _____

19. Determine the population after 1 year and after 2 years.

19. _____

20. Determine the population after 3 years and after 4 years.

20. _____

For #21–24, some say local newspaper subscriptions are decreasing by 5% per year, and the rate will remain the same for the next 5 years.

21. Determine the decay factor.

21. _____

22. The number of subscriptions now is 200,000. If the decay rate remains constant at 5% per year, write an algebraic rule to determine the number of subscriptions after t years.

22. _____

23. Determine the number of subscriptions after 1 year and after 2 years.

23. _____

24. Determine the number of subscriptions after 4 years and after 5 years. Round to the nearest subscription.

24. _____

For #25–28, you invest in a 5-year certificate of deposit that earns 3% per year, and the rate remains the same for the next 5 years.

25. Determine the growth factor.

25. _____

26. You buy the CD for $5000. If the interest rate remains constant at 3% per year, write an algebraic rule to determine the value after *t* years.

26. _____

27. Determine the value of the CD after 1 year and after 2 years.

27. _____

28. Determine the value of the CD after 3 years and after 4 years. Round to the nearest cent.

28. _____

Concept Connections

29. How do you know if an exponential function has a growth or decay factor?

_____

_____

30. With an exponential function, where is the independent variable located?

_____

_____

Chapter 4 AN INTRODUCTION TO NONLINEAR PROBLEM SOLVING

Activity 4.9

Learning Objectives
1. Recognize the equivalent forms of the direct variation statement.
2. Determine the constant of proportionality in a direct variation problem.
3. Solve direct variation problems.

Practice Exercises

1. J varies directly with d. Write an equation relating J, d and   **1.** _____
the constant of variation, k.

2. From Exercise #1, if $J = 13$ when $d = 52$, determine the   **2.** _____
value of k.

3. From Exercises #1 and #2, rewrite the equation from   **3.** _____
Exercise #1 using the value of k from Exercise #2.

4. From Exercises #1, #2, and #3, determine J when $d = 16$.   **4.** _____

5. M varies directly with p. Write an equation relating M, p and the constant of variation, k.

5. _____

6. From Exercise #5, if $M = 150$ when $p = 40$, determine the value of k.

6. _____

7. From Exercises #5 and #6, rewrite the equation of Exercise #5 using the value of k from Exercise #6.

7. _____

8. From Exercises #5, #6, and #7, determine M when $p = 90$.

8. _____

9. z varies directly as the square of q. Write an equation relating z, q and the constant of variation, k.

9. _____

10. From Exercise #9, if $z = 81$ when $q = 6$, determine the value of k.

10. _____

11. From Exercises #9 and #10, rewrite the equation from   **11.** _____
Exercise #9 using the value of k in Exercise #10.

12. From Exercises #9, #10, and #11, determine z when $q = 7$.   **12.** _____

13. V varies directly as the cube of r. Write an equation   **13.** _____
relating V, r, and the constant of variation, k.

14. From Exercise #13, if $V = 384$ when $r = 4$, determine the   **14.** _____
value of k.

15. From Exercises #13 and #14, rewrite the equation of   **15.** _____
Exercise #13 using the value of k from Exercise #14.

16. From Exercises #13, #14, and #15, determine V when   **16.** _____
$r = 10$.

17. *y* varies directly as *x*, *y* = 30 when *x* = 15. Determine *y* when *x* = 11.

17. _____

18. *y* varies directly as *x*, *y* = 9 when *x* = 63. Determine *y* when *x* = 28.

18. _____

19. *y* varies directly as *x*, *y* = 90 when *x* = 1.5. Determine *x* when *y* = 225.

19. _____

20. *y* varies directly as x^2, *y* = 48 when *x* = 4. Determine *y* when *x* = 17.

20. _____

21. *y* varies directly as x^2, *y* = 175 when *x* = 5. Determine *x* when *y* = 1008.

21. _____

22. y varies directly as x^2, $y = 72$ when $x = 6$. Determine y when $x = 5$.

22. _____

23. y varies directly as x^3, $y = 320$ when $x = 4$. Determine y when $x = 7$.

23. _____

24. y varies directly as x^3, $y = 20$ when $x = 2$. Determine x when $y = 540$.

24. _____

25. y varies directly as x^3, $y = 189$ when $x = 3$. Determine y when $x = 6$.

25. _____

26. y varies directly as x, $y = 52$ when $x = 4$. Determine x when $y = 91$.

26. _____

27. y varies directly as x^3, $y = 28$ when $x = 2$. Determine x when $y = 224$.

27. _____

28. y varies directly as x^2, $y = 272$ when $x = 4$. Determine x when $y = 425$.

28. _____

Concept Connections

29. Name other equivalent ways to state y varies directly as x.

_____

_____

30. You are creating a one-tenth model of your favorite classic car. You have several choices when ordering the specialized pieces from an online vendor. If the original size of the bumper is 7.5 feet, which size bumper do you purchase: 7.5-inch piece, 9-inch piece, or 10-inch piece? Hint: convert feet to inches first.

_____

_____

Chapter 4 AN INTRODUCTION TO NONLINEAR PROBLEM SOLVING

Activity 4.10

Learning Objectives
1. Recognize functions of the form $y = \frac{k}{x}$, $x \neq 0$, as nonlinear.
2. Recognize equations of the form $xy = k$ as inverse variation.
3. Graph an inverse variation relationship from symbolic rules.
4. Solve equations of the form $\frac{a}{x} = b$, $x \neq 0$.

Practice Exercises

For #1-14, solve for x.

1. $\dfrac{60}{x} = 20$ **2.** $\dfrac{90}{x} = 15$ **1.** _____

                                                                    2. _____

3. $\dfrac{22}{x} = 55$ **4.** $\dfrac{144}{x} = 96$ **3.** _____

                                                                    4. _____

5. $180 = \dfrac{150}{x}$ **6.** $13 = \dfrac{65}{x}$ **5.** _____

                                                                    6. _____

7. $-17 = \dfrac{153}{x}$

8. $-18 = \dfrac{81}{x}$

9. $-21 = \dfrac{-15}{x}$

10. $35 = \dfrac{105}{x-3}$

11. $4 = \dfrac{16}{x+3}$

12. $25 = \dfrac{900}{x+12}$

13. $175 = \dfrac{280}{x-2}$

14. $15 = \dfrac{210}{x-3}$

15. Use the equation $y = \dfrac{7}{x}$ to determine y when $x = 21$.

15. _____

16. Use the equation $y = \dfrac{7}{x}$ to determine x when $y = 35$.

16. _____

17. Use the equation $y = \dfrac{0.2}{x}$ to determine y when $x = 100$.

17. _____

18. Use the equation $y = \dfrac{0.2}{x}$ to determine x when $y = 20$.

18. _____

19. Use the equation $xy = 20$ to write y as a function of x.

19. _____

20. Use the equation $y = \dfrac{15}{x}$ to determine y when $x = 1.5$.

20. _____

21. Use the equation $y = \dfrac{15}{x}$ to determine x when $y = 90$.

21. _____

22. Use the equation $xy = 3.2$ to write y as a function of x.

22. _____

23. Use the equation $y = \dfrac{8}{x}$ to determine y when $x = \dfrac{1}{3}$.　　　**23.** _____

24. Use the equation $y = \dfrac{8}{x}$ to determine x when $y = 6.4$.　　　**24.** _____

25. Use the equation $xy = 0.8$ to write y as a function of x.　　　**25.** _____

26. The current I, varies inversely as the resistance R. The equation $6 = IR$ represents the relationship of a 6-volt circuit. Rewrite the equation to express the current, I, as a function of the resistance.　　　**26.** _____

27. From Exercise #26, if the resistance is 3 ohms, what is the current, in amps?　　　**27.** _____

28. From Exercise #26, if the current is 12 amps, what is the resistance, in ohms?　　　**28.** _____

Concept Connections

29. For the general function, $y = \dfrac{k}{x}$, what is the domain?

_____

_____

30. Why is an equation of the form $xy = k$ known as inverse variation?

_____

_____

Chapter 4 AN INTRODUCTION TO NONLINEAR PROBLEM SOLVING

Activity 4.11

Learning Objectives
1. Recognize functions of the form $y = a\sqrt{x}$ as nonlinear.
2. Evaluate and solve equations the involve square roots.
3. Graph square root functions from numerical data.
4. Graph square root functions from symbolic rules.

Practice Exercises
For #1-28, solve for x.

1. $\sqrt{x} = 7$

2. $\sqrt{x} + 4 = 9$

3. $\sqrt{x} = 13$

4. $\sqrt{x} - 5 = 11$

5. $\sqrt{x} - 6 = 14$

6. $5\sqrt{x} = 60$

1. _____

2. _____

3. _____

4. _____

5. _____

6. _____

7. $16\sqrt{x} = 400$

8. $7\sqrt{x} = 28$

7. _____

8. _____

9. $\sqrt{x+1} = 6$

10. $\sqrt{x+4} = 7$

9. _____

10. _____

11. $\sqrt{x-3} = 8$

12. $\sqrt{4x} = 6$

11. _____

12. _____

13. $\sqrt{7x} = 21$

14. $\sqrt{6x} - 5 = 0$

13. _____

14. _____

15. $\sqrt{9x} - 8 = 1$

16. $\sqrt{12x} - 10 = 14$

15. _____

16. _____

Name:

Instructor:

Date:

Section:

17. $4x^2 = 144$

18. $5x^2 = 245$

17. _____

18. _____

19. $6x^2 + 1 = 97$

20. $x^2 + 11 = 0$

19. _____

20. _____

21. $7x^2 + 1 = 253$

22. $8x^2 - 5 = 7x^2 + 20$

21. _____

22. _____

23. $5x^2 - 6 = 3x^2 + 26$

24. $9x^2 + 3 = 579$

23. _____

24. _____

25. $6x^2 + 11 = 5x^2 + 15$

26. $8x^2 + 6 = 5x^2 + 81$

25. _____

26. _____

27. $x^2 + 6 = 2$ **28.** $11x^2 - 111 = 7x^2 + 85$ **27.** _____

28. _____

Concept Connections

29. When defining a general square root function $y = a\sqrt{x}$, it is stated that x is nonnegative. Why?

_____

_____

30. Does the graph of the general square root function $y = a\sqrt{x}$ include $x = 0$? If not, why not? If so, what is the corresponding value for y?

_____

_____

Odd Answers

Chapter 1 NUMBER SENSE

Activity 1.3

Key Terms
1. commutative
3. formula
5. variable

Practice Exercises
7. 518
9. 6
11. 25
13. 150
15. 36
17. 8.567×10^{10}
19. 3,049,900,000,000
21. 1.8×10^7, or 18,000,000
23. $A = 66$
25. $P = 32$
27. $V = 450$

Concept Connections
29. Melanie didn't perform the addition inside the parentheses first.
   The solution is $3 \cdot (8 + 7) \cdot 5 = 3 \cdot 15 \cdot 5 = 45 \cdot 5 = 225$.

Activity 1.4

Practice Exercises
1. $8\dfrac{11}{12}$
3. $6\dfrac{22}{35}$
5. $10\dfrac{17}{30}$
7. $7\dfrac{1}{12}$
9. $\dfrac{8}{15}$
11. $2\dfrac{23}{30}$
13. $\dfrac{1}{10}$
15. $21\dfrac{1}{9}$
17. $88\dfrac{2}{5}$
19. 2
21. $\dfrac{27}{32}$
23. $\dfrac{7}{15}$
25. $3\dfrac{11}{36}$
27. $6\dfrac{2}{3}$

Concept Connections

29. Alice did not find a common denominator before adding the numerators.

The solution is $\dfrac{2}{3}+\dfrac{7}{9}=\dfrac{6}{9}+\dfrac{7}{9}=\dfrac{13}{9}=1\dfrac{4}{9}$.

Activity 1.5

Key Terms

1. weighted average

Practice Exercises

3. 85
5. 88
7. 70
9. 85
11. 8
13. 7
15. 6
17. 83
19. 71
21. 74
23. 79
25. 0.346
27. 0.419

Concept Connections

29. Each score has the same weight of $\dfrac{1}{5}$.

Activity 1.6

Key Terms

1. ratio
3. verbal
5. relative measure

Practice Exercises

7. $\dfrac{55}{103}$
9. $\dfrac{13}{20}$
11. $\dfrac{99}{100}$
13. 0.85
15. 0.25
17. 0.467
19. 74%
21. 43.75%
23. 12.5%
25. 0.045
27. 0.0063

Concept Connections

29. The missing information is the number of baskets attempted for each player.

Activity 1.7

Key Terms
1. divided by
3. proportional reasoning

Practice Exercises
5. 32
7. 24
9. 81
11. 8
13. 800
15. 40
17. 324
19. 19,000
21. 1400
23. 96.6
25. 17,500
27. 3048

Concept Connections
29. There are 408 right-handed children.

Activity 1.8

Key Terms
1. relative change

Practice Exercises
3. $15
5. $24
7. $20
9. history
11. 3.2%
13. 39.1%
15. 35 lbs
17. 35 lbs
19. 6
21. 13
23. lack of water
25. 6.5%
27. 6.1%

Concept Connections
29. The percent decrease is 75%. The amount of decrease cannot exceed the original amount.

Activity 1.9

Key Terms
1. decay factor

Practice Exercises
3. 1.0875
5. $1.75
7. 1.041
9. 24,598
11. 2.15%
13. $55,200
15. 16.7%
17. 5.4%
19. $119.99
21. 200 pounds

23. 20.25 mg   25. $167.58

27. $52.49

Concept Connections

29. The original value is larger since the decay factor is less than 1.

Activity 1.10

Key Terms

1. product

Practice Exercises

3. 0.75   5. $95.63

7. 0.90   9. $56,782.69

11. 20 lbs   13. $2772

15. 0.275   17. 0.21

19. $373,763   21. $84,248

23. 0.75

25. Decay factor, since the value is less than 1.

27. 6.25% decrease

Concept Connections

29. The cumulative effect for the first deal is 49% decrease. The cumulative effect for the second deal is 44% decrease. The first deal is better.

Activity 1.11

Practice Exercises

1. 54,000 seconds   3. 2.485 miles

5. 70.87 inches   7. 25,200 minutes

9. 1.44 grams/30 days   11. 4.5 yards

13. 190,080 inches   15. 0.42 grams/12 weeks

17. 138,336 feet   19. 42,155.8 meters

21. 10.6 kpL   23. 2.11 qt

25. 600 ft   27. 504 hours

Concept Connections

29. All the French recipes used Metric system, and Julia had to convert everything to American units for her cookbook. Since she didn't use conversion rate analysis, she had to replicate each recipe through trial and error, which was very time consuming.

Activity 1.12

Practice Exercises

1. −15

3. −12

5. −7

7. −19

9. −18

11. 9

13. 0

15. −11

17. −5

19. −3

21. $-\dfrac{4}{7}$

23. 15

25. <

27. >

Concept Connections

29. Golf is scored using par for each hole, with the number of shots over (positive) or under (negative) par.

Activity 1.13

Practice Exercises

1. 48

3. −21

5. −3.14

7. 0

9. 28

11. 5

13. undefined

15. −6

17. 72

19. $-\dfrac{1}{16}$

21. −12.4

23. −0.2

25. $\dfrac{3}{5}$

27. undefined

Concept Connections

29. The change in height is $\dfrac{-6 \text{ feet}}{6 \text{ days}}$, or −1 foot per day.

Activity 1.14

Practice Exercises

1. 7

3. −80

5. $\dfrac{19}{15}$

7. −1

9. 0

11. -1.953125×10^{2}

13. 64

15. $\dfrac{1}{64}$

17. −25

19. −15

21. 0

23. 4

25. –4

27. 0.0042

Concept Connections

29. $4^4 = 256$ and $4^{-4} = \dfrac{1}{256}$

Chapter 2 VARIABLE SENSE

Activity 2.1

Key Terms
1. input; output

3. variable

5. horizontal

Practice Exercises
7. Vertical axis

9. Horizontal axis

11. y

13. 2003

15. 2002 through 2005

17. 2005 through 2006

19. No; having a negative number of frequent flier miles doesn't make any sense.
21. (Answers will vary.) 0, 1, 2, ….1,000,000.
23. No, the lowest point in California is Death Valley, which is –282 ft.
25. Yes; Mount Whitney has elevation 14,494 ft.
27. (Answer will vary.) Elevations from –282 ft to 14,494 ft would be reasonable.

Concept Connections
29. Each unit of time is equal and is expressed as 1 year.

Activity 2.2

Key Terms
1. scaling

3. quadrants

5. output

Practice Exercises
7.

9. x-axis

11. first quadrant

13. third quadrant

15. first quadrant

17. third quadrant

19. *y*-axis

21. fourth quadrant

23. (–4, 3)

25. (3, –2)

27. (5, 0)

Concept Connections

29. On the first graph, the points form a straight line. On the second graph they form a zig zag pattern, depending on the spacing between tick marks.

Activity 2.3

Key Terms

1. replacement

Practice Exercises

3.

Input	2	3	4	5	6	7
Output	6	9	12	15	18	21

5.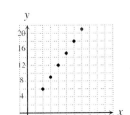

7. Add 5 to the input to obtain the output.

9.

Input	−3	0	3	6	9	12
Output	−12	0	12	24	36	48

11.

Input	−3	−1	1	3	5	7
Output	−8	−6	−4	−2	0	2

13.

Input	−4	−2	0	2	4	8
Output	2	4	6	8	10	12

15.

Input	−4	−2	0	2	4	6
Output	−2	−1	0	1	2	3

17.

Input	−3	−1	1	3	6	7
Output	9	1	1	9	36	49

19.

Input	Output is 3 less than the input
0	−3
3	0
6	3
9	6
12	9

21.

Input	Output is 2 times the input plus 7
−4	−1
−3	1
−2	3
−1	5
0	7
1	9

23. $6

25.

Pounds of grapes	2	3	6	8	10
Cost	$1.50	$4.50	$9.00	$12.00	$15.00

27.

Input	Output is 5 times the input minus 4
−3	−19
−2	−14
−1	−9
0	−4
1	1
2	6

Concept Connections

29. No, you can't buy a negative quantity of grapes.

Activity 2.4

Key Terms

1. coefficient

3. terms

Practice Exercises

5. $x + 8$

7. $x - 12$

9. $7x - 4$

11. $9 + 5x$

13. −11

15. 15

17. 23

19. −15.5

21. 23

23. −4

25. $22

27. $5d$

29. (Answers will vary.) You can use any variable letter, however the most common are: p for pennies, n for nickels, d for dimes, q for quarters; $0.01p + 0.05n + 0.10d + 0.25q$

Activity 2.6

Key Terms
1. subtract
3. add
5. solution

Practice Exercises
7. $y = 81$
9. $z = 165$
11. $y = -175.2$
13. $y = -8$
15. $y = 3.4$
17. $z = -28$
19. $y = 213.9$
21. $z = 324$
23. $y = -4$
25. $y = 1.6$
27. $y = -316.8$

Concept Connections
29. The total amount charged is $34.50.

Activity 2.7

Key Terms
1. reverse; inverse
3. subtract; multiply

Practice Exercises
5. $x = 5$
7. $x = -4$
9. $x = 0$
11. $x = 20$
13. $x = 10$
15. $x = 3$
17. $x = 12$
19. $x = 7$

21.

x	y
7	26
13	56

23.

x	y
$\frac{3}{4}$	22
8	-7

25.

x	y
36	3
90	12

27.

x	y
2	-17
-4	-29

Concept Connections
29. Division is only performed with nonzero numbers.

Activity 2.8

Practice Exercises

1. $d = \dfrac{c}{8}$

3. $d = 9$

5. $y = 5$

7. $p = 3A - q - r$

9. $a = 24$

11. $r^2 = \dfrac{A}{4\pi}$

13. $r = 5$

15. $c = 7\pi$

17. $s = \dfrac{P}{4}$

19. $s = 12$

21. $z = \dfrac{xy}{t}$

23. $x = \dfrac{P - b}{a}$

25. $n = \dfrac{G - w}{150}$

27. $h = \dfrac{V}{lw}$

Concept Connections

29. (2) It is more efficient to substitute for y and then solve for x, since there are fewer steps to reach the answer.

Activity 2.9

Practice Exercises

1. $x = 24$

3. $x = 30$

5. $x = 54$

7. $x = 522$

9. $x = 1600$

11. $x = 195$

13. 374 miles

15. 84,000 miles

17. 875 shares

19. 105 games

21. False

23. True

25. False

27. False

Concept Connections

29. Bill is right. After cross multiplication, both proportions are equivalent to $a \cdot d = b \cdot c$.

Activity 2.10

Practice Exercises

1. $y = x - 9$

3. $y = 8 + \dfrac{x}{3}$

5. $y = -5x - 16$

7. $y = -8(x + 11)$

9. $y = 6x - 17$

11.

x	$x \cdot 4$	$4x$
-3	-12	-12
-1	-4	-4
0	0	0
2	8	8

13. Commutative property of multiplication

15. No, the expressions are not equivalent.

x	$x - 5$	$5 - x$
-6	-11	11
-1	-6	6
0	-5	5
1	-4	4
6	1	-1

17. No

19.

x	$y_1 = 4x - 1$	$y_2 = 4(x - 1)$
-2	-9	-12
0	-1	-4
2	7	4
3	11	8

21. No. They generate different outputs.

23. y_3: start with $x \rightarrow$ square $\rightarrow$ subtract $1 \rightarrow$ to obtain y.

y_4: start with $x \rightarrow$ subtract $1 \rightarrow$ square $\rightarrow$ to obtain y.

25.

x	$y_5 = \dfrac{1}{2}x^2 + 1$	$y_6 = \dfrac{1}{2}(x^2 + 2)$
-2	3	3
0	1	1
4	9	9
6	19	19

27. Yes. They generate identical outputs.

Concept Connections

29. $49 \cdot 3 = (50 - 1) \cdot 3 = 150 - 3 = 147$

Activity 2.11

Key Terms

1. term

3. common

Practice Exercises

5. $30x - 10$

7. $24.5 - 10.5x$

9. $-9x + 11y + z$

11. $-\dfrac{3}{13}x + \dfrac{1}{4}$

13. $3(5x - 4y)$

15. $7x(3x - 5)$

17. $10 - x$

19. $36 - 24x - 28y$

21. $10x + 24$

23. $-x^2 + 20x$

25. $4x - 27$

27. $16x - 55$

Concept Connections

29. $5y^2$ and $3y^2$ are like terms because they contain identical variable factors, including exponents. $3y$ does not have the same exponent as the other two terms.

Activity 2.12

Practice Exercises

1. $-5n + 18$

3. $14x - y + 41$

5. $-14x - 51$

7. $x - 8$

9. $x - 9$

11. $12x - 5$

13. $x + 7$

15. $1 - 7n$

17. $24x - 9$

19. $2x - 1$

21. $x - 4$

23. $3x - 6$

25. $25x - 83$

27. $8x + 216$

Concept Connections

29. Answers will vary. Some examples are GPS directions and recipes.

Activity 2.13

Practice Exercises

1. $x = 6$

3. $x = 4$

5. $x = 97$

7. $t = 3$

9. $x = -27$

11. $x = -3$

13. No solution

15. $x = 11$

17. $m = \dfrac{y - b}{x}$

19. $y = \dfrac{4x - 6}{3}$

21. $r = \dfrac{A - P}{Pt}$

23. $s = \dfrac{P - 2b}{2}$

25. $x = \dfrac{6y + 30}{5}$

27. $y = 18x - 20$

Concept Connections

29. The equation is an identity. The answer is all real numbers.

Chapter 3 FUNCTION SENSE AND LINEAR FUNCTIONS

Activity 3.1

Practice Exercises

1. Yes	3. No
5. Yes	7. Yes
9. Yes	11. Yes
13. No	15. No
17. Yes	19. No
21. Decreasing	23. Decreasing
25. Increasing	27. Increasing

Concept Connections

29. A function is a rule relating an input variable and an output variable in a way that assigns one and only one output to each input value.

Activity 3.2

Practice Exercises

1. $f(5) = 14$	3. $h(-3) = 39$
5. $f(-4) = -26$	7. $h(3) = 17$
9. $g(3) = -3$	11. $x = 2$
13. $t = 72$	15. $a = -3$
17. $x = 8$	19. $w = 15$

21. $\{-4, -2, 0, 2, 4, 6\}$

23. f is increasing over the set $\{0, 2, 4, 6\}$.

25. The minimum value of f is –2, and it occurs when x is 0.

27. $f(-4) = 6$

Concept Connections

29. The collection of all possible input values is called the domain of the function. In functions that result from contextual situations, the domain consists of input values that make sense within the context. Such a domain is often called a practical domain.

Activity 3.3

Key Terms

1. remains constant	3. decreases

Practice Exercises

5. $\Delta y = 1$	7. Increasing
9. $\Delta y = -4$	11. Decreasing

13. $\Delta w = 10$

15. Increasing

17. $\Delta y = 0$

19. Constant

21. $\dfrac{\Delta y}{\Delta x} = \dfrac{2}{3}$

23. $\dfrac{\Delta y}{\Delta x} = \dfrac{11}{10}$

25. $\dfrac{\Delta y}{\Delta x} = -6$

27. $\dfrac{\Delta y}{\Delta x} = \dfrac{8}{9}$

Concept Connections

29. average rate of change = $\dfrac{\text{change in output}}{\text{change in input}}$

Activity 3.4

Practice Exercises

1. $\dfrac{\Delta \text{output}}{\Delta \text{input}} = \dfrac{14}{7} = \dfrac{14}{7} = \dfrac{28}{14} = 2$; linear

3. $\dfrac{\Delta \text{output}}{\Delta \text{input}} = \dfrac{5}{5} = \dfrac{5}{5} = \dfrac{15}{15} = 1$; linear

5. $\dfrac{\Delta \text{output}}{\Delta \text{input}} = \dfrac{29.6}{4} = \dfrac{22.2}{3} = \dfrac{14.8}{2} = 7.4$; linear

7. $(-2, 14)$ and $(-10, 8)$

9. $(4, 6)$ and $(2, -2)$

11. $(5, 8)$ and $(-5, 4)$

13.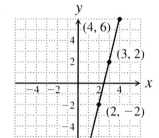

15. $m = 5$

17. $m = \dfrac{5}{2}$

19. $m = 3$

21. $m = 1$

23. $m = \dfrac{3}{4}$

25. 6.25%

27. 5%

Concept Connections

29. slope $= m = \dfrac{y_2 - y_1}{x_2 - x_1}$, where (x_1, y_1) and (x_2, y_2) are any two points on the line.

Activity 3.5

Key Terms

1. vertical

3. first

Practice Exercises

5. $m = 0$

7. $m = -3$

9. $m = -\dfrac{1}{3}$

11. $m =$ undefined

13. $m = -\dfrac{1}{6}$

15. $(0, 0)$

17. $(3, 0)$

19. $(-3, 0)$

21. $(0, 0)$

23. $(0, -3)$

25. $(3, 6), (0, 6), (-5, 6)$

27. $(2, -1)$ and $(-4, 3)$

Concept Connections

29. It started to freeze about 7:30 PM.

Activity 3.6

Practice Exercises

1. slope: 5; y-intercept: $(0, 2)$

3. slope: $\dfrac{1}{3}$; y-intercept: $(0, -8)$

5. slope: 0; y-intercept: $(0, 25)$

7. slope: -7; y-intercept: $(0, 210)$

9. slope: -8; q-intercept: $(0, 64)$

11. $y = 3x - 2$

13. $y = \dfrac{4}{5}x + 7$

15. $y = -\dfrac{2}{5}x$

17. $y = 0$

19. $y = -5x - 5$

21. Yes

23. Yes

25. No

27. No

Concept Connections

29. $y = mx + b$

Activity 3.7

Practice Exercises

1. slope: 4; y-intercept $(0, -3)$
   x-intercept: $\left(\frac{3}{4}, 0\right)$, or $(0.75, 0)$

3. slope: -1; y-intercept $(0, 3)$
   x-intercept: $(3, 0)$

5. slope: $+\frac{1}{2}$; y-intercept $(0, 1)$
   x-intercept: $(2, 0)$

7. slope: $-\frac{4}{3}$; y-intercept $(0, 4)$
   x-intercept: $(3, 0)$

9. $y = -2x + 9$

11. $y = \dfrac{5}{2}x - 5$

13. $m = 3$

15. $m = -1$

17. $m = 0$

19. slope is undefined

21.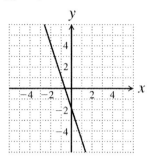

23.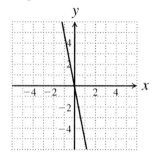

25.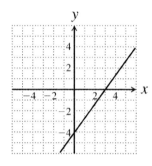

27.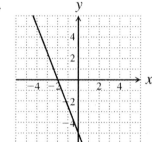

Concept Connections

29. When a line is written in slope-intercept form, the slope and y-intercept can be determined without any calculation.

Activity 3.8

Practice Exercises

1. slope: 2; y-intercept $(0, 5)$

3. slope: $\dfrac{5}{2}$; y-intercept $(0, -6)$

5. slope: -1; y-intercept $(0, 7)$

7. slope: 0; y-intercept $(0, 8)$

9. slope: $\dfrac{1}{5}$; y-intercept $\left(0, \dfrac{1}{2}\right)$

11. slope: -1; y-intercept $(0, 2)$

13. slope: 0; y-intercept $(0, 3)$

15. slope: 1; y-intercept $(0, 0)$

17. slope: 3; y-intercept $(0, -2)$

19. slope: 2; y-intercept $(0, 9)$

21. slope: $-\dfrac{1}{2}$; y-intercept $(0, -5)$

23. slope: -1.2; y-intercept $(0, 4)$

25. slope: -3; y-intercept $(0, 1)$

27. slope: $\dfrac{5}{3}$; y-intercept $(0, -3)$

Concept Connections

29. $m = \dfrac{y_2 - y_1}{x_2 - x_1} = \dfrac{-1 + 3}{6 - 4} = \dfrac{2}{2} = 1$ and $m = \dfrac{y_2 - y_1}{x_2 - x_1} = \dfrac{-3 + 1}{4 - 6} = \dfrac{-2}{-2} = 1$.

Activity 3.9

Practice Exercises

1. $y = 2x - 5$

3. $y = \dfrac{1}{3}x - 3$

5. $y = 5$

7. $y = 6$

9. slope: -3; y-intercept: $(0, -3)$

11. slope: $\dfrac{1}{3}$; y-intercept: $(0, -1)$

13. slope: 4; y-intercept: $(0, 2)$

15. slope: 1; y-intercept $(0, -3)$

17. $y = \dfrac{4}{5}x - 2$

19. $y = 6$

21. $y = -\dfrac{13}{6}x$

23. $y = -4$

25. $y = -2x + 17$

27. $y = -3x + 2$

Concept Connections

29. Point-slope form is $y - y_1 = m(x - x_1)$.

Activity 3.12

Key Terms

1. consistent

Practice Exercises

3. $x = 2$, $y = 3$

5. Using a graphing calculator, we have

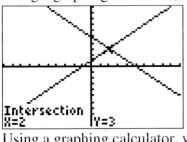

7. $x = -3$, $y = -12$

9. Using a graphing calculator, we have

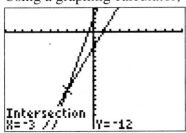

11. $x = 0$, $y = -2$

13. Using a graphing calculator, we have

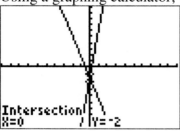

15. no solution, the system is inconsistent

17. Using a graphing calculator, we have

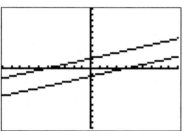

19. $x = \dfrac{1}{2}$, $y = -1$

21. Using a graphing calculator, we have

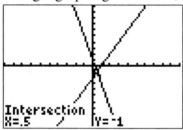

23. $x = 1$, $y = 5$

25. Using a graphing calculator, we have

27. $x = 2$, $y = 1$

Concept Connections

29. It is possible that an incorrect answer will check in the first equation, but not in the second. To fully check an answer, Tom must use both equations in his check.

Activity 3.13

Practice Exercises

1. $x = 1$, $y = 1$

3. $x = 0$, $y = -2$

5. $x = 1.5$, $y = 2.5$

7. no solution

9. $x = -1.4$, $y = 3$

11. $x = 7$, $y = 4$

13. $x = -8$, $y = -5$

15. $x = 5.5$, $y = 2.5$

17. $x = -2$, $y = 4$

19. $x = 3$, $y = 1$

21. $x = 1$, $y = -7$

23. $x = -4$, $y = 1$

25. $x = 6$, $y = 2$

27. $x = -2$, $y = 0$

Concept Connections

29. Each of the lines has the same slope but different y-intercepts. Therefore the lines are parallel and the system has no solution.

Activity 3.15

Practice Exercises

1. $x > 25$

3. $x \geq 21$

5. $x < \dfrac{3}{10}$

7. $x > -\dfrac{7}{4}$

9. $x < -3$

11. $x \geq 9$

13. $x > 32$

15. $x \leq -8$

17. $x \leq -5$

19. $x \geq 11$

21. $x > -1$

23. $x \geq 9$

25. $x > 4$

27. $x > 0$

Concept Connections

29. For $x \geq 2$, the 2 is part of the solution. For $x > 2$, the 2 is not in the solution.

Chapter 4 AN INTRODUCTION TO NONLINEAR PROBLEM SOLVING

Activity 4.1

Key Terms

1. degree

3. monomial

5. standard form

Practice Exercises

7. $4.2a + 8.1b$

9. $0.8x - 18$

11. $S(2.5) = 16.25$

13. $p(3.7) = 5.6$

15. $9x^3 - 5x^2 + 6x + 9$

17. $-5x^3 + 10x^2 - 11x$

19. $-2x^2 + 4x + 1$

21. $5x^3 - 3x^2 + 13x - 13$

23. -14

25. degree: 3; $-7x^3 + 19x + 15$

27. degree: 6; $4x^6 + 19x^3 - 3x + 14$

Concept Connections

29. Paul is right. $\sqrt{3} + x$ is a polynomial, but $\sqrt{3 + x}$ is not.

Activity 4.2

Practice Exercises

1. $4x^6$

3. $-48w^{10}$

5. $6x^{15}$

7. $-8.4x^{16}$

9. $-3s^8t^9$

11. $\dfrac{2}{5w^2}$, or $\dfrac{2}{5}w^{-2}$

13. $\dfrac{125a^9}{27b^6}$

15. $12x^3$

17. $\dfrac{1}{x^4}$, or x^{-4}

19. $5x^2 + 35x$

21. $3x^5 + 4x^4 - x^3$

23. $20t^8 - 12t^6 - 4.8t^3$

25. $21x^5 - 98x^4$

27. $45x^5 - 20x^4 - 12x^2$

Concept Connections

29. Answers will vary. Remind Jeff what $\left(x^3\right)^4$ means: $\left(x^3\right)^4 = x^3 \cdot x^3 \cdot x^3 = x^{3+3+3} = x^9$. Also, substituting a value for y (other than 0 or 1), will help Jeff check his answer: $y^5y^3 = y^{5+3} = y^8$, let $y = 2$. $2^52^3 = 32 \cdot 8 = 256$, and $2^8 = 256$.

Activity 4.3

Key Terms

1. difference of squares

Practice Exercises

3. $x^2 + 10x + 16$

5. $x^2 + 3x - 40$

7. $x^2 - 16$

9. $12x^2 + 13x + 3$

11. $4x^2 - 19x + 12$

13. $6a^2 - 13ab + 6b^2$

15. $x^3 - x^2 - 6x + 18$

17. $2x^2 - x^2y - 15xy + 5xy^2 + 25y^2$

19. $x^3 - 64$

21. $x^2 - 10x + 25$

23. $25x^2 - 16$

25. $x^2 - 50x + 625$

27. $x^2 - 22x + 121$

Concept Connections

29. Conjugate binomials are binomials whose first terms are identical and whose second terms differ only in sign.

Activity 4.4

Practice Exercises

1. $x = \pm 4$

3. $x = \pm 6$

5. $x = \pm\sqrt{15} \approx \pm 3.87$

7. $c = \pm 7$

9. $t = \pm 8$

11. $x = 2$ or $x = -8$

13. $x = 15$ or $x = -5$

15. $x = 6$ or $x = -2$

17. $x = 4$ or $x = -\dfrac{4}{3}$

19. $x = 0$ or $x = -\dfrac{11}{2}$

21. $x = \dfrac{11}{7}$ or $x = -1$

23. $x = 1$ or $x = -19$

25. $x = 19$ or $x = -11$

27. $x = 9$ or $x = -15$

Concept Connections

29. A quadratic function has a U-shaped curve. The solutions are where the curve crosses the x-axis.

Activity 4.5

Practice Exercises

1. $x = 0$ or $x = -15$

3. $x = 4$ or $x = -5$

5. $x = \dfrac{1}{9}$ or $x = -\dfrac{5}{2}$

7. $x = 0$ or $x = 5$

9. $x = 0$ or $x = 7$

11. $x = 0$ or $x = 11$

13. $x = 0$ or $x = \dfrac{5}{4}$

15. $x = 0$ or $x = -25$

17. $y = 0$ or $y = -6$

19. $x = 6$ or $x = -2$

21. $p = 0$ or $p = -1$

23. $p = 0$ or $p = -1$

25. $x = \dfrac{1}{4}$ or $x = -\dfrac{1}{3}$

27. $x = 0$ or $x = 18$

Concept Connections

29. The zero-product property: If the product of two factors is zero, then at least one of the factors must also be zero. Stated symbolically, if $a \cdot b = 0$, then either $a = 0$ or $b = 0$.

Activity 4.6

Practice Exercises

1. $x = -1$ or $x = -7$

3. $x = 5$ or $x = -4$

5. $x = 6$ or $x = 7$

7. $y = -7$ or $y = -3$

9. $x = 6$ or $x = 3$

11. $x = 4$ or $x = -2$

13. $x = 9$ or $x = -3$

15. $x = 7$ or $x = -5$

17. $x = 3$ or $x = -13$

19. $x = 8$ or $x = 5$

21. $x = 8$ or $x = 6$

23. $x = 6$ or $x = -1$

25. $x = 1$ or $x = 13$

27. $x = 3$ or $x = -4$

Concept Connections

29. The general form of a quadratic equation is $y = ax^2 + bx + c$, $a \neq 0$.

Activity 4.7

Practice Exercises

1. $x = \dfrac{1}{3}$ or $x = 1$

3. $x = -1.5$ or $x = -8.5$

5. $x = 0$ or $x = 2.75$

7. $x = 0$ or $x = -1.8$

9. $x = 0.75$ or $x = 3$

11. $x = 9.82$ or $x = -0.82$

13. $x = 5.24$ or $x = 0.76$

15. $x = 6.21$ or $x = 1.29$

17. $x = 0.79$ or $x = -0.36$

19. $x = 4.57$ or $x = -6.57$

21. $x = -15$ or $x = -9$

23. $x = -34$ or $x = 4$

25. $x = -\dfrac{5}{3}$ or $x = -\dfrac{3}{2}$

27. $x = -\dfrac{5}{3}$ or $x = 1$

Concept Connections

29. The quadratic formula is $x = \dfrac{-b \pm \sqrt{b^2 - 4ac}}{2a}$.

Activity 4.8

Practice Exercises

1. 1.07

3. 1 day: 53,500 bugs;
    2 days: 57,245 bugs

5. 1.40

7. 1 year: 322 million texts;
    2 years: 451 million texts

9. 0.80

11. 1 year: $800
    2 years: $640

13. 0.65

15. 1 hour: 19,500 bugs;
    2 hours: 12,675 bugs

17. 1.10

19. 1 year: 165,000;
    2 years: 181,500

21. 0.95

23. 1 year: 190,000;
    2 years: 180,500

25. 1.03

27. 1 year: $5150.00;
    2 years: 5304.50

Concept Connections

29. If the base b is greater than 1, then it has a growth factor. If the base b is between 0 and 1, then it has a decay factor.

Activity 4.9

Practice Exercises

1. $J = kd$

3. $J = 0.25d$

5. $M = kp$

7. $M = 3.75p$

9. $z = kq^2$

11. $z = 2.25q^2$

13. $V = kr^3$

15. $V = 6r^3$

17. $y = 22$

19. $x = 3.75$

21. $x = 12$

23. $y = 1715$

25. $y = 1512$

27. $x = 4$

Concept Connections

29. Answers may vary. Other statements may include: y is directly proportional to x; $y = kx$ for some constant k; The ratio $\dfrac{y}{x}$ always has the same constant value, k.

Activity 4.10

Practice Exercises

1. $x = 3$

3. $x = 0.4$

5. $x = \dfrac{5}{6}$

7. $x = -9$

9. $x = \dfrac{5}{7}$

11. $x = 1$

13. $x = 3.6$

15. $y = \dfrac{1}{3}$

17. $y = 0.002$

19. $y = \dfrac{20}{x}$

21. $x = \dfrac{1}{6}$

23. $y = 24$

25. $y = \dfrac{0.8}{x}$

27. $I = 2$ amps

Concept Connections

29. The domain is all real numbers except $x = 0$.

Activity 4.11

Practice Exercises

1. $x = 49$

3. $x = 169$

5. $x = 400$

7. $x = 625$

9. $x = 35$

11. $x = 67$

13. $x = 63$

15. $x = 9$

17. $x = \pm 6$

19. $x = \pm 4$

21. $x = \pm 6$

23. $x = \pm 4$

25. $x = \pm 2$

27. no real solution

Concept Connections

29. The square root of a negative number is not real.